ADVANCED BIOSIGNAL PROCESSING AND DIAGNOSTIC METHODS

Edited by **Christoph Hintermüller**

Advanced Biosignal Processing and Diagnostic Methods

http://dx.doi.org/10.5772/61919
Edited by Christoph Hintermüller

Contributors

Jorge Muñoz-Minjares, Oscar Ibarra Manzano, Yuriy S. Shmaliy, Radek Martinek, Radana Kahankova, Hana Skutova, Jaromir Konecny, Petr Bilik, Jan Zidek, Homer Nazeran, Dany Obeid, Sarah Samad, Sawsan Sadek, Gheorghe Zaharia, Ghaïs El Zein, Jesus Pastor, Ancor Sanz, Lorena Vega-Zelaya, Guillermo Ortega, Rafael G Sola, Ching Te Chiu, Yu Sheng Chan, Wen Te Liu

CBS, Edition **2017**

Published by InTech

Janeza Trdine 9, 51000 Rijeka, Croatia

Publishing Process Manager Dajana Pemac

Technical Editor SPi Global

Cover Designer InTech Design Team

Additional hard copies can be obtained from orders@intechopen.com

Advanced Biosignal Processing and Diagnostic Methods, Edited by Christoph Hintermüller
p. cm.
Print ISBN 978-953-51-2519-8
Online ISBN 978-953-51-2520-4

INTECH

open science | open minds

Contents

Preface

Personal health and well-being was and is important for all individuals. This includes the way people are living, what they do to stay healthy as well as a profound, well-informed diagnosis and appropriate treatment in case of disease. To achieve these goals, modern medicine is provided with a large variety of tools to assess a patient's health state and collect the information required for a proper diagnosis and treatment, which is tailored to the patient's needs.

Many of these available tools use signals either generated by the human body, for example, electroencephalogram (EEG) and electrocardiogram (ECG), or by interacting with the human body while traversing it like microwaves or reflected visible light that is recorded by a video camera. The biosignals recorded by the available and newly developed methods have to be processed to extract the information about the patient's condition and analyzed tissue and cells. This book presents some recent developments in the field of biosignal processing.

The first chapter presents a method that allows to non-intrusively assess the respiratory state of a patient just by observing his walking pattern. The proposed marker-less camera-based system can be easily integrated into clinical diagnostic process. The second chapter takes the concept of contactless assessment of the patient's health state even a step further. It uses microwave radar signals for contactless recording respiratory and cardiovascular parameters. The advantage of microwave signals casted at the patient's body surface is that they can travel several meters and thereby even penetrate thick layers of stones and concrete. Therefore, one of the applications of these methods envisioned by the authors is the monitoring of vital signs of people buried under their house after an earthquake while the rescue team is trying to remove the blocking obstacles and to prepare the necessary treatment to stabilize the subjects' condition.

Thanks to modern medicine, children who are born too early or with severe conditions have a good prognosis in surviving their first years and leading a healthy and normal life, thanks to continuously evolving and improving prenatal diagnostics. This includes the recording of the fetal ECG used to assess its cardiovascular state either through interuterine recordings or by abdominal recordings of the mother. The first method is highly invasive and therefore poses high risks for the mother and the fetus. Therefore, the authors of the third chapter propose an advanced method to record the fetal ECG (fECG) from the mother's abdomen. The proposed non-linear approach is compared with existing ones with respect to their capability to provide additional cardiovascular parameters, such as PQ, QT, and ST time of the fetal ECG. The main challenge is to properly filter out the maternal ECG (mECG), which overlaps with the fECG, the abdominal muscles, and intestinal signals, while preserving the small signal of the fECG.

Epilepsy, which has to be considered a severe neurological condition, is characterized by recurrent seizures. In the past, it was assumed that these epileptic events are generated by a single focus and spread from there to other brain regions. Recent findings in neurological and epilepsy research indicate that a seizure is rather an abnormal activity and response of the whole network neuronal network involved. Chapter 4 presents a new network analysis approach, which uses neurophysiological data recorded prior to surgical epilepsy treatment to identify the affected brain regions and their interconnections activated during an epileptic event.

In the twentieth century, the DNA was identified to control all functions of individual cells, tissues, organs, and the whole body. Even though already available since the past century only very recently reasonably fast and accurate methods to read the DNA and analyze its structure have been developed. With this development, it became possible to use the information stored in the DNA for diagnostic purpose. Alterations in the structure of the DNA, for example, introduced when it is copied to generate proteins or when it is repaired after one of its strands or both broke, can severely affect the function of the cell. A large number of so-called copy number alterations (CNA) can be used to characterize cancer cells. A new method to assess and validate the CNA in the DNA of single cells is presented in Chapter 5.

This book only covers a small selection of new and advanced biosignal processing and diagnostic methods. As an editor, I thank all the authors for their contributions. I also thank all researchers whose work could not be included in this book for various reasons. All your great work is an important contribution to the field of biosignal processing with respect to the development of new and improved diagnostic tools, therapies, and treatments.

Dr. Christoph Hintermüller
Guger Technologies OG
Schiedlberg, Austria

Chapter 1

Classifying and Predicting Respiratory Function Based on Gait Analysis

Yu Sheng Chan, Wen Te Liu and Ching Te Chiu

Additional information is available at the end of the chapter

http://dx.doi.org/10.5772/63917

Abstract

The human walking behaviour can express the physiological information of human body, and gait analysis methods can be used to access the human body condition. In addition, the respiratory parameters from pulmonary spirometer are the standard of accessing the body condition of the subjects. Therefore, we want to show the correlation between gait analysis method and the respiratory parameters. We propose a vision sensor-based gait analysis method without wearing any sensors. Our method proposed features such as D'_p, V'_p and γv to prove the correlation by classification and prediction experiment. In our experiment, the subjects are divided into three levels depending on the respiratory index. We run classifying and predicting experiment with the extracted features: V'_p and γv. In the classifying experiment, the accuracy result is 75%. In predicting experiment, the correlations of predicting the forced expiratory volume in 1 s (FEV1) and forced vital capacity (FVC) are 0.69 and 0.67, respectively. Therefore, there is a correlation between the pulmonary spirometer and our method. The radar system is a tool using impulse to record the moving of the subjects' chest. Combining the features of radar system with our features improves the classification result from 75 to 81%. In predicting FEV1/FVC, the correlation also improves from 25 to 42%. Therefore, cooperating with radar system improves the correlation.

Keywords: gait analysis, classification, prediction, pattern recognition, feature extraction

1. Introduction

Walking behaviour can express information of human body-like pathological symptoms. For example, Parkinson's disease patients are characterized by special pace rhythm [1].

People increase the respiratory ventilation when they are walking or exercising. However, those people who suffer from chronic obstructive airway disease (COAD) cannot increase their respiratory ventilation quick enough to maintain the exercising behaviour. Consequently, they change their behaviour such as walking slowly so that they can maintain their respiratory ventilation. We can perform gait analysis on COAD patients because their walking behaviours are different with normal people when they are exercising because of their respiratory function.

Chronic obstructive pulmonary disease (COPD) is one condition of COAD. Nowadays, there are many chronic diseases in our daily life. COPD is one of them. COPD is a chronic airway disease characterized by progressive going downhill of the breathing functions [2]. One characteristic of COPD disease is the decreasing of forced expiratory volume in 1 s (FEV1) because of the obstructive airway [2]. Depending on their disease severity, they have different walking behaviours. Therefore, gait analysis can be used to judge the COAD patients' airway condition by observing their walking behaviour. However, it is difficult to collect the data of COAD patients without medical staff. Without clinical data, we cannot verify the correctness of our gait analysis algorithm.

By cooperating with Shuang-Ho Hospital in New Taipei, Taiwan, we set up an experiment. We film the side view of the subjects when they are performing a 6-min brisk walking test. By gait analysis, we can extract the features from walking behaviour such as pace distance and walking speed variation. However, in order to obtain that physiological information, we may need to wear sensors or markers on the subjects. Our method does not need to wear any sensors on the subjects.

In gait analysis, it is common to wear markers or sensors to record walking behaviour. In the experiment [1], the subjects need to wear a recorder on the ankle so that it could record the stride interval. In another experiment [3], the subjects also need to place a designed insole with 12 sensors into their own shoes.

However, there are some drawbacks of using makers or sensors. Firstly, it is inconvenient and uncomfortable of attaching them on human body and might affect the normal walking. Secondly, some sensors are heavy or hard to use for the elderly. Thirdly, some sensors have the electromagnetic interference that might affect and harm human body. In addition to the sensor problem, it is hard to tell a subject suffering from COAD disease or not by a single experiment. Without a complete examination, it is hard to judge whether the subjects are COAD patients or not. Consequently, we decide to access the respiratory function. By the pulmonary spirometer, we can obtain the tested subjects' respiratory data. Nowadays, the parameters from pulmonary spirometer are the standard to access the respiratory function.

2. Related work

2.1. Chronic obstructive pulmonary disease

COPD is one condition of COAD and we introduce some studies of COPD disease. The research of the World Health Organization (WHO) and the global initiative for chronic obstructive lung disease revealed that nowadays the COPD is the fourth cause of death in the United States. For the entire world, the COPD can be considered the fifth cause of death [4]. In another study, the COPD can be considered the fourth leading cause of death in the world [5] and is projected to be the fifth by 2020 as a worldwide burden of disease [6].

Generally speaking, chronic disease is gradually rising because of the ageing population and changing habits. The treatment for COPD patients is pulmonary rehabilitation programmes, including when patients are discharged at home [7]. Economic analyses have shown that over 70% of COPD-related health-care expenditures result from emergency room visits and hospital care for exacerbations; this translates to >$10 billion annually in the Unites States [8]. COPD patients have a lower physical activity level than healthy peers [9]. The reduced level of physical activity also related to an increased risk [10]. Liao et al. [11] proposed a review that concentrated on describing wearable devices for measuring physical activity level in COPD patients. In [12], the authors evaluated a method for detecting an exacerbation onset in COPD patients. They used data collected through a pulse oximeter, permitting an easy way to the cohort of patients composed of elderly people affected by COPD. The study [13] provided a system, which offers an effective platform for the satisfaction of the clinical and the patient's needs in the area of early diagnosis of patient's health status. The portable system aims at the effective management in the health status of the patients who are suffering from COPD.

2.2. Gait analysis

Gait analysis plays an important role in accessing human's walking behaviour and it aims to extract biomechanics information. The obvious disease on lower limbs is Parkinson's disorder. In [14], they provided a feasible method image marker to measure gait with little skin movement. Then, they performed quantitative analysis to extract gait parameter. According to their analysis, the joint angle, rotation angle of lower limbs, stride velocity and stride length have significant difference between Parkinson's disease patients and non-diseased subjects. The study [1] demonstrated that the gait variability in terms of statistical parameters of stride interval such as STC_{γ}, would be increased in Parkinson's disorder. In addition to Parkinson's disease, Alzheimer's disease (AD) also can be detected by gait analysis. The study [15] presents an inertial-sensor-based wearable device and its associated stride detection algorithm to analyse gait information for patients with Alzheimer's disease.

The above methods all use markers or sensors to get interest points and then use gait analysis. With the advances in smart phones, Susu Jiang et al. [16] used a smart phone with an accelerometer and a gyroscope to collect human walking gait data in daily life. However, one smart phone only records one feature data. If we want to acquire many interest point data, testers need to tie many smart phones on the body, which might be inconvenient for walking.

Now, we proposed a gait analysis algorithm without any wearable marker or sensor. In addition, we also collect clinical COAD and control data to verify the correlation between pulmonary spirometer and our gait analysis method.

3. Vision sensor-based gait analysis

To avoid wearing sensors on the tested subjects, we propose a vision sensor-based gait analysis method. This method is composed of four parts: input video frame decomposition, pre-processing, feature extraction and gait analysis. The input video takes the side view of the subjects when they are performing a 6-min brisk walking test. Then, the input video is decomposed into individual frames for further processing.

In the pre-processing part, there are three components to extract the silhouette of the tested subject: background subtraction, shadow removal and connected component labelling (CCL). The background subtraction subtracts the background image by the current video frame to capture the moving object. Because the human shadow cannot be removed by background subtraction, we adopt Jamie and Richard's method [17] to solve the shadow problem. Then, there are some noises that cannot be removed, so we use the connected component-labelling method to reduce the noises and obtain the maximum object.

In the feature extraction part, there are two steps to obtain the desired features: segmentation and feature extraction. In the segmentation part, we have to find the central of gravity (COG) first, and use the COG point to perform body segmentation and extract legs of the subjects. Then, we can get the features such as pace distance and pace velocity in the feature extraction part.

In the gait analysis part, we divide whole subjects into two groups as the Bad and Good by consulting the proposed respiratory index formula. We use support vector machine (SVM) to perform classification and use adaptive network-based fuzzy inference system (ANFIS) to perform prediction.

3.1. Pre-processing

3.1.1. Background subtraction

We take the first frame of the input video as the background image. The background subtraction method is shown in Eq. (1). The components x and y are the pixel location. The factor t is the current frame number. I represents the RGB value of the pixel, which is located at (x, y) and F is the -subtraction result. In our experiment, we set the Th-value at 15

$$F = \left| I(x,y,t) - B(x,y,1) \right| > Th \tag{1}$$

3.1.2. Shadow removal

After subtract background image, some interferences still exist. The result of background subtraction is shown in **Figure 1**. The human shadow is viewed as foreground and we need to remove it. We consult Jamie and Richard's [17] method to solve the shadow problem. The method is divided into two parts: (1) brightness distortion and (2) chromatic distortion.

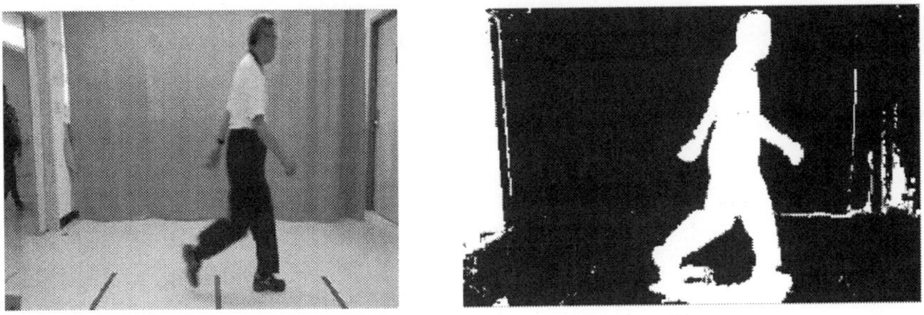

Figure 1. (a) Input frame and (b) background subtraction result.

I_i is the ith pixel of the input frame which can be represented in RGB space by the vector I^i = [(IR(i), IG(i))] as shown in **Figure 2**. E^i is the ith pixel of the background image which can be represented as E^i = [(ER(i), EG(i), EB(i))]. The lengths of these lines are the intensity of the ith pixel. The projection of I_i onto E_i is denoted as $\alpha_i E_i$. We call the α_i as brightness distortion. We can solve α_i by Eq. (2)

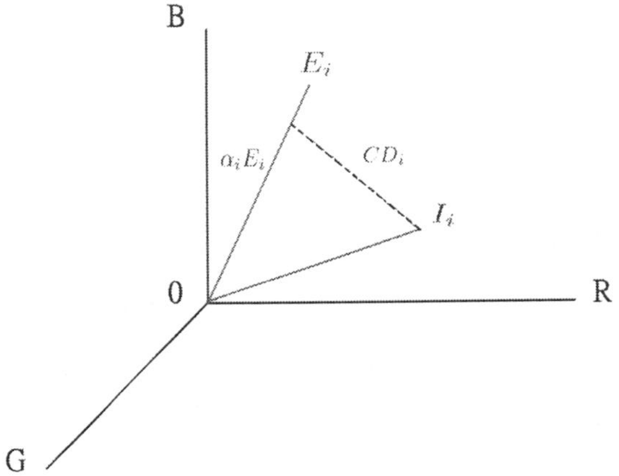

Figure 2. Colour representation in RGB space.

$$\alpha_i = \min_{\alpha_i} \left\| \acute{I}_i - \alpha_i \acute{E}_i \right\|^2 \tag{2}$$

$$\alpha_i < \tau_{BD} : Foreground \tag{3}$$

There is a threshold τ_{BD}. We take those pixels whose α_i values are smaller than threshold τ_{BD} as foreground such as Eq. (3) expressing. In our experiment, we set the threshold τ_{BD} at 0.7.

In the chromatic distortion part, we calculate the distance in RGB space between I_i and E_i. **Figure 2** shows this as line CD_i and we can solve CD_i value by Eq. (4)

$$CD_i = \left\| \acute{I}_i - \alpha_i \acute{E}_i \right\| \tag{4}$$

In the same way, we also set a threshold τ_{CD} to determine this pixel as background or foreground such as Eqs. (5) and (6) expressing. Those pixels whose CD_i values are greater than threshold τ_{CD} are viewed as foreground and those pixels whose values are smaller as background. In our experiment, we set the threshold τ_{CD} at 10.

$$CD_i > \tau_{CD} : Foreground \tag{5}$$

$$Otherwise : Background \tag{6}$$

After finishing these two parts above, we combine these two images and the background subtraction result together to have a result without shadow.

3.1.3. Connected component labelling

Connected component labelling is used to detect the connected components and label them. These components have their own label number. **Figure 3** shows an example. **Figure 3(a)** shows three different regions and **(b)** shows the labels of the regions. We keep the largest group as our result. In this case, the region of label 3 is reserved and discards other regions.

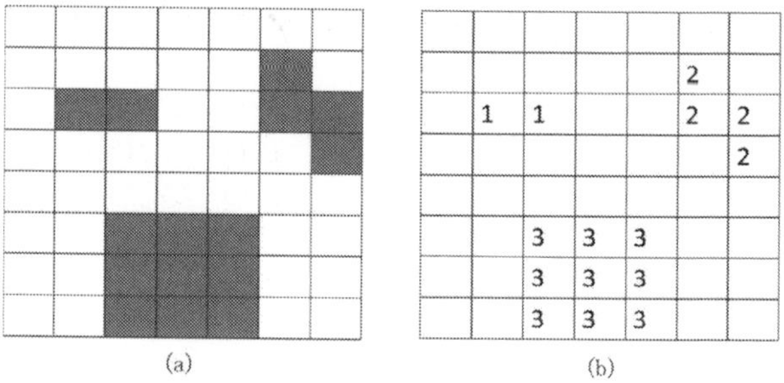

Figure 3. (a) Three disconnected components and (b) labelling image.

Though we have an image without shadow, some noises exist. We perform connected component labelling method so that we can have an image without these noises. **Figure 4** shows the flow of pre-processing.

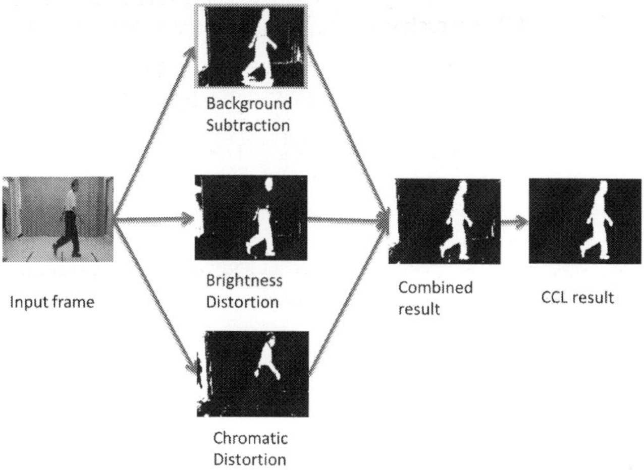

Figure 4. The flow of pre-processing.

3.2. Feature extraction

After the pre-processing, we get the complete target silhouette. In this section, we separate legs from extracted silhouette by segmentation. Then, we find out the gait features such as pace distance and pace velocity in the feature extraction part.

3.2.1. Segmentation

In the segmentation part, we need to find the central of gravity first. Then, we use the extracted silhouette to get the contour by edge detection in order to build the distance map (DM). We can separate legs from human silhouette by DM.

We find the central of gravity from the extracted human silhouette by Eq. (7). After finding the COG of the whole body (COG_x, COG_y), we use edge detection on human silhouette to extract human contour such as in **Figure 5**. Then, we draw a DM by computing the Euclidean distance between (COG_x, COG_y) and extracted human contour shown in **Figure 6(b)**. We compute the distance map by Eq. (8) and x_i, y_i are the location of the ith pixel of the extracted human contour.

$$(COG_x, COG_y) = \left(\frac{\sum_{i=1}^{N} Body_{xi}}{N}, \frac{\sum_{i=1}^{N} Body_{yi}}{N}\right) \tag{7}$$

$$DM = \sqrt{(COG_x - x_i)^2 + (COG_y - y_i)^2} \tag{8}$$

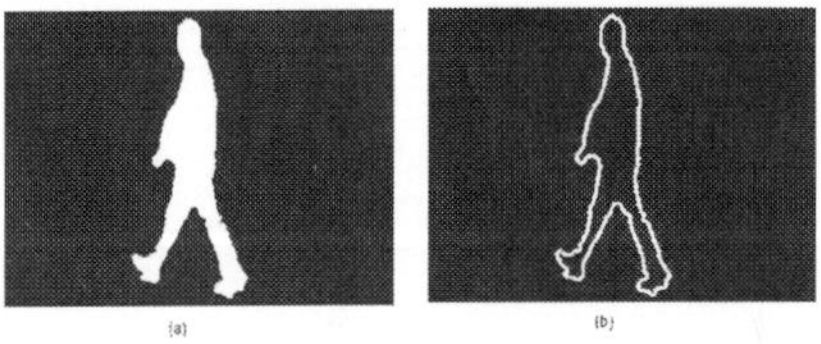

Figure 5. (a) Silhouette image and (b) the contour image of (a).

We find the Nlegs, Nb2(l), Nb2(r) three points from DM as shown in **Figure 6**. Connecting Nb2(l) and Nb2(r) divides human body into top and down two parts. Connecting (COG_x, COG_y) and Nlegs separates legs into leg(l) and leg(r) shown in **Figure 7**.

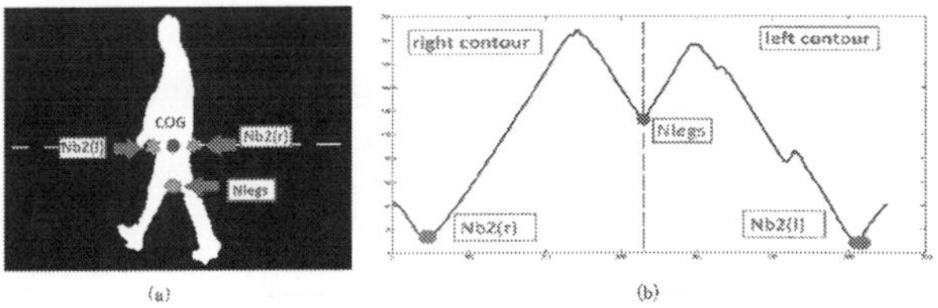

Figure 6. (a) Finding COG(x, y) in silhouette. (b) Distance map of COG(x, y).

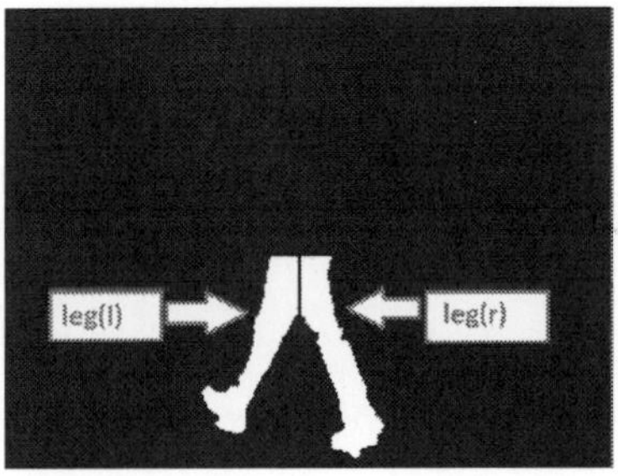

Figure 7. Separated legs.

3.2.2. Gait features

In this part, we extract gait features such as pace distance, pace time and pace velocity. **Figure 8** shows a pace cycle model and we can extract pace distance (D_p), pace time (T_p) and pace velocity (V_p) from this model and Eq. (9). The distance value of the pace model comes from the horizontal distance between leg(l) and leg(r) such as D1. If the feet are close, we calculate the distance of the closed feet such as D2. D1 and D2 are the longest and shortest distance values, respectively. T1 and T2 are the start and end frame numbers of this step, respectively, so we have to multiply the frame rate to get the actual T_p. The frame rate is 1/30 s per frame in our experiment. Then, V_p comes from dividing D_p by T_p.

$$\begin{cases} D_p = D1 - D2 \\ T_p = (T2 - T1) * framerate \\ V_p = D_p / T_p \end{cases} \tag{9}$$

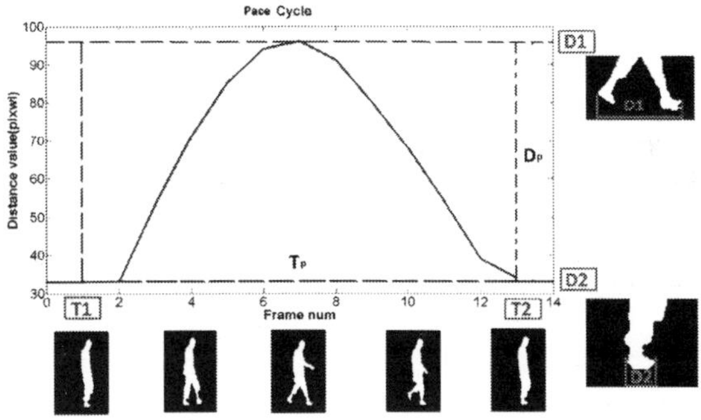

Figure 8. The pace cycle model.

3.3. Gait analysis

To verify that there exists a correlation between pulmonary spirometer and our system, we perform classification and prediction experiments. There are two groups, namely bad and good, which are classified by the parameters from pulmonary spirometer and it becomes our classification standard. We take support vector machine for classification and take adaptive network-based fuzzy inference system for prediction. Here, we introduce the tools: SVM and ANFIS.

3.3.1. Support vector machine

SVM comes from Vapnik's statistical learning theory [18] and it is a machine-learning method, which can be a powerful tool for learning from data and for solving classification problem [18]. In a two-group classification problem such as our study (Bad/Good), the target is to find the *Hyperplane* between the two data groups. SVM finds the *Hyperplane* by looking for the maximum margin between two groups. The main idea of SVM is to transform data into higher dimensions and then construct a *Hyperplane* between two classes in the transformed space. Those data vectors, which are nearest to the constructed line in the transformed space, are called the support vectors which contain information about *Hyperplane*. **Figure 9** shows the concept about the SVM.

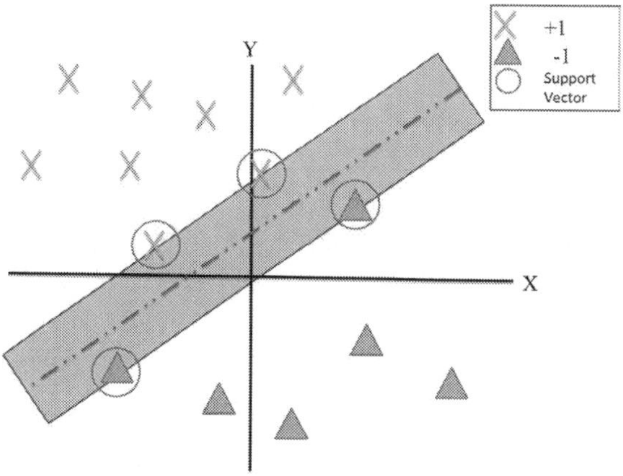

Figure 9. Example of two-group problem showing optimal *Hyperplane* (dotted line).

3.3.2. Adaptive network-based fuzzy inference system

ANFIS was presented by Jang in 1993 [19]. Adaptive network-based fuzzy inference system can construct an input-output mapping based on human knowledge by a hybrid-learning algorithm. The fuzzy inference system is employed with adaptive network. ANFIS contains a five-layer forward neural network to construct the inference system.

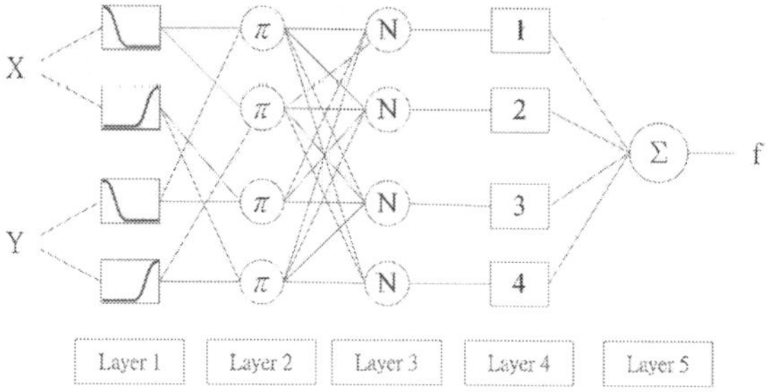

Figure 10. ANFIS structure with two inputs and four rules.

Input space is mapped to a given membership function (MF). By membership function, the input becomes a degree between 0 and 1. With different membership functions and the number of membership functions, the results are different. **Figure 10** shows the ANFIS structure with two inputs and four rules.

http://dx.doi.org/10.5772/63917

The study [20] explained the function of each layer. In Layer$_{(1)}$, the outputs are the membership function degree of the inputs, which are given by Eqs. (10) and (11)

$$O_{1,i} = \mu_{Ai}(x), i = 1,2 \tag{10}$$

$$O_{1,i} = \mu_{Bi-2}(y), i = 3,4 \tag{11}$$

where x and y are the inputs to node i.

Layer$_{(2)}$ involves fuzzy operations. ANFIS fuzzifies the inputs by using AND operation. The label _ means that they perform simple multiplier. Equations (12) and (13) can show the output of Layer$_{(2)}$

$$O_{2,i} = w_i = \mu_{A1}(x) * \mu_{Bi}(y), i = 1,2 \tag{12}$$

$$O_{2,i} = w_i = \mu_{A2}(x) * \mu_{Bi-2}(y), i = 3,4 \tag{13}$$

In Layer$_{(3)}$, the label N indicates normalization. This layer can be represented by Eq. (14)

$$O_{3,i} = \acute{w}_i = \frac{w_i}{w_1+w_2+w_3+w_4}, i = 1,2,3,4 \tag{14}$$

Layer$_{(4)}$ is the product of the normalized data which can be represented as Eq. (15). The parameters p_i, q_i and r_i are determined during the training process

$$O_{4,i} = \acute{w}_i f_i = \acute{w}_i (p_i x + q_i x + r_i), i = 1,2,3,4 \tag{15}$$

Layer$_{(5)}$ implements sum of all inputs such as Eq. (16)

$$O_{5,i} = \sum \acute{w} f_i = \frac{\sigma_i w_i f_i}{\sigma_i w_i} \tag{16}$$

4. Proposed gait features

We propose some gait features from D_p and V_p and call the ith step of D_p and V_p as D_{pi} and V_{pi}. The mean distance and mean velocity are denoted as $D_p^{'}$ and V_p, respectively. In addition, we divide all steps into S sections depending on the step counts in order to reveal the variation of the subject's movement during the 6-min brisk walking test. **Figure 11** shows an example that we divide D_p of all steps into six sections and there are μ_{D_i} and σ_{D_i} of each section.

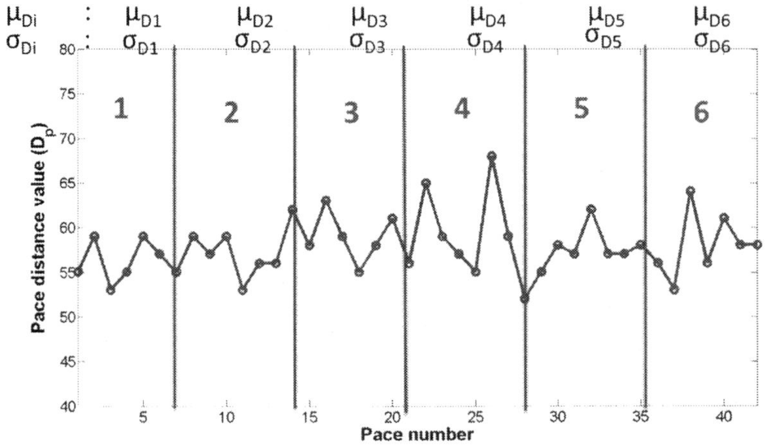

Figure 11. The six sections of D_p.

The mean of distance for section i is called as μ_{D_i} and the mean of velocity for section i is called as μ_{V_i}. The variance of distance for section i is called as σ_{D_i} and the variance of velocity for section i is called as σ_{V_i}. These parameters are listed in Eq. (17)

$$\begin{cases} N: Totalstepcounts, n = \frac{N}{S}, \\ \acute{D_p} = \frac{1}{N} * \sum_{i=1}^{N} D_{pi}, \acute{V_p} = \frac{1}{N} * \sum_{i=1}^{N} V_{pi} \\ \mu_{D_i} = \frac{1}{n} * \sum_{j=(i-1)*n+1}^{(i)*n} D_{pj}, \mu_{V_i} = \frac{1}{n} * \sum_{j=(i-1)*n+1}^{(i)*n} V_{pj} \\ \sigma_{D_i} = \sqrt{\frac{1}{n} * \sum_{j=(i-1)*n+1}^{(i)*n} \left(D_{pj} - \mu_{D_i}\right)^2}, \sigma_{V_i} = \sqrt{\frac{1}{n} * \sum_{j=(i-1)*n+1}^{(i)*n} \left(V_{pj} - \mu_{V_i}\right)^2} \end{cases} \tag{17}$$

From σ_{D_i} and σ_{V_i}, we can calculate the mean of distance variance and the mean of velocity variance, which are denoted as $\sigma'_{D(1,S)}$ and $\sigma'_{V(1,S)}$. The mean of distance variance for the first two sections and the mean of velocity variance for the first two sections are denoted as $\sigma'_{D(1,2)}$ and $\sigma'_{V(1,2)}$, respectively. The mean of distance variance for the last two sections and the mean of velocity variance for the last two sections are denoted as $\sigma'_{D(S-1,S)}$ and $\sigma'_{V(S-1,S)}$ respectively. In addition, we also calculate the distance variance ratio and velocity variance ratio, which are denoted as γ_D and γ_V. The distance variance ratio is defined as the multiplication of $\sigma'_{D(1,S)}$ and the result of dividing $\sigma'_{D(S-1,S)}$ by $\sigma'_{D(1,2)}$. The velocity variance ratio is defined as the multiplication of $\sigma'_{V(1,S)}$ and the result of dividing $\sigma'_{V(S-1,S)}$ by $\sigma'_{V(1,2)}$. **Figure 12** shows the region of these parameters and these parameters are listed in Eq. (18)

$$\begin{cases} \sigma_{D(\acute{1},S)} = \frac{1}{S} * \sum_{i=1}^{S} \sigma_{D_i}, \sigma_{V(\acute{1},S)} = \frac{1}{S} * \sum_{i=1}^{S} \sigma_{V_i} \\ \sigma_{D(\acute{1},2)} = \frac{1}{2} * \sum_{i=1}^{2} \sigma_{D_i}, \sigma_{V(\acute{1},2)} = \frac{1}{2} * \sum_{i=1}^{2} \sigma_{V_i} \\ \sigma_{D(\acute{S}-1,S)} = \frac{1}{2} * \sum_{i=S-1}^{S} \sigma_{D_i}, \sigma_{V(\acute{S}-1,S)} = \frac{1}{2} * \sum_{i=S-1}^{S} \sigma_{V_i} \\ \gamma_D = \frac{\sigma_{D(\acute{1},S)} * \sigma_{D(\acute{S}-1,S)}}{\sigma_{D(\acute{1},2)}}, \gamma_V = \frac{\sigma_{V(\acute{1},S)} * \sigma_{V(\acute{S}-1,S)}}{\sigma_{V(\acute{1},2)}} \end{cases} \tag{18}$$

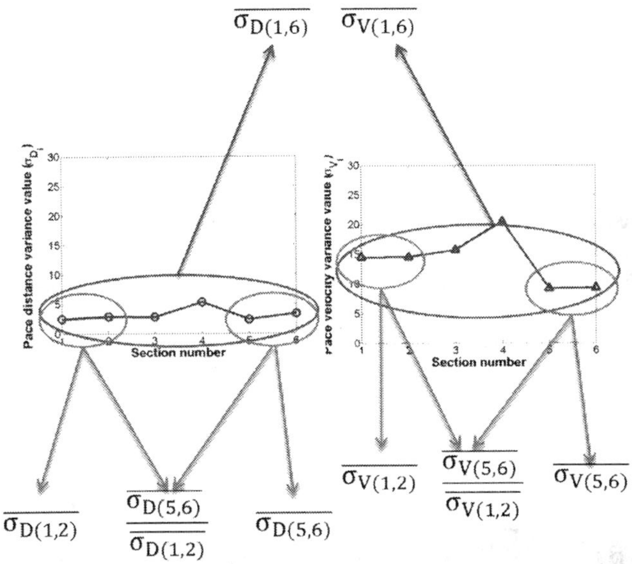

Figure 12. The variance mean of whole sections (purple region) and first two sections (orange region) and last two sections (green region).

5. Clinical experiment environment

5.1. Experiment set-up and flow

The experiments are run at Shuang-Ho Hospital in New Taipei, Taiwan. We film the side view of the subjects when they perform the 6-min brisk walking test. We set up a green curtain to exclude the interferences such as the movement of other people from our experiment. We film the walking subjects using Nikon P330 digital camera.

Firstly, the therapists ask the subjects' profile including height, weight and age. Secondly, by using pulmonary spirometer, we can get respiratory parameters such as FEV1 and FVC data about the subjects. Thirdly, before the experiment starts, the therapist helps the subjects wear a pulse oximeter on the index finger, which is used to measure the oxygen and pulse. Fourthly, the subjects need to take a 2-min break so that the pulse oximeter can record the oxygen and pulse in normal condition. Fifthly, when the walking test begins, the subjects should walk along the trail as fast as possible. While the subjects start their walking test, we film the side view of the subjects. Sixthly, after the 6-min walking test, the subjects use the pulmonary spirometer to measure the respiration parameter after exercising again.

5.2. Data collection

We run the experiments from September 2014 to July 2015. In the experiments, there are 60 subjects who aged between 24 and 91 years. Among these 60 subjects, there are 48 men and 12 women.

There are two rooms: the subjects walk from the right room to the left one, then turnaround to walk into the right room. When the subjects walk to the other side of the trail, they need to turnaround and continue to walk along the trail. They decrease their walking speed so that they can turnaround easily when they are close to the border. To avoid recording those slowdown steps, we abandon those steps and keep normal steps. Taking **Figure 13** as an example, there are six steps in this walking trail. We just consider steps 1 and 6 as the normal steps and abandon steps 2–5.

Depending on the respiratory index that comes from Eq. (19), these subjects are divided into three levels: level 1 (the worst respiratory function), level 2 (poor respiratory function) and level 3 (normal respiratory function). **Table 1** shows the respiratory index used to classify the three levels. We call the respiratory index as *REX*. The smaller *REX* represents the worse respiratory function

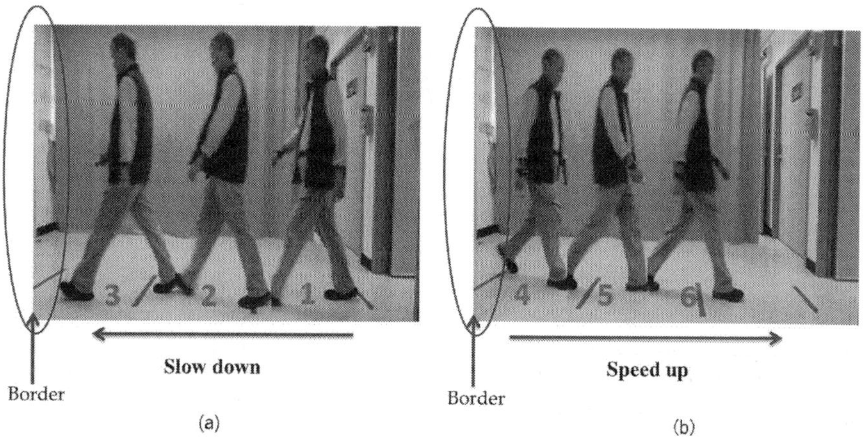

Figure 13. (a) Walking trail before turnaround. (b)Walking trail after turnaround.

$$REX = postFEV\frac{\frac{\frac{1*postFEV1}{preFEV1}*postIC}{preIC}*postFVC}{preFVC}*post\frac{FEV1}{FVC} \tag{19}$$

Group	***Level* 1**	***Level* 2**	***Level* 3**
REX	0.7	[0.7,1.65]	1.65

Table 1. The respiratory index (*REX*) used to classify the three levels.

The main item of REX formula is postFEV1 and other items are used to adjust it. The three items of $\frac{\text{postFEV1}}{\text{preFEV1}}$, $\frac{\text{postIC}}{\text{preIC}}$ and $\frac{\text{postFVC}}{\text{preFVC}}$ are greater than one in normal respiratory function subjects but smaller than one in poor respiratory function subjects. The value of post$\frac{\text{FEV1}}{\text{FVC}}$ is lower than 0.75 in those subjects who have poor respiratory function. **Figure 14** shows the lung capacity changes of respiratory factors.

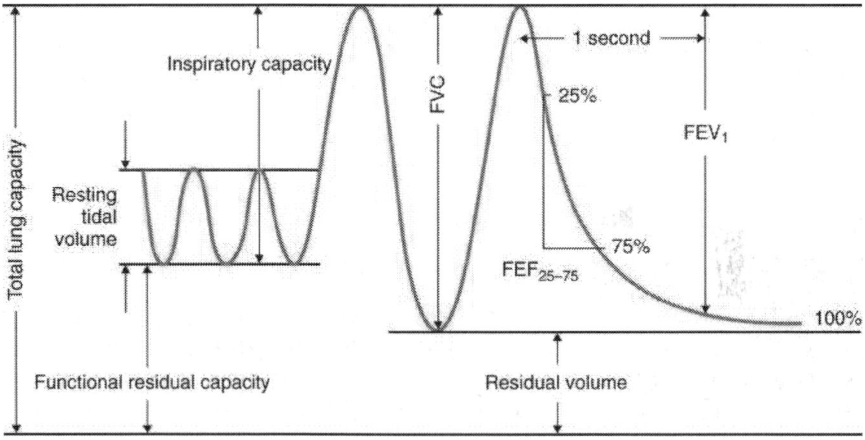

Figure 14. Lung capacity changes [21].

The FEV1 is the volume that has been exhaled at the end of the first second of forced expiration. The FVC is the forced vital capacity that is used for the determination of the vital capacity from a maximally forced expiratory effort. The IC is the inspiratory capacity that is the sum of inspiratory reserve volume and tidal volume. The $\frac{\text{FEV1}}{\text{FVC}}$ is the ratio that is used for the diagnosis of obstructive and restrictive lung disease.

6. Experimental results

In this chapter, we use the features that we get from Chapter 4 to perform our experiment. There are three experiments: classification with support vector machine, prediction with adaptive neural fuzzy inference systems and cooperating with radar system.

In the first experiment, we choose one feature from D_p', V_p' and the other feature is γ_V. There are two combinations. Among these two combinations, we find the best one according to the classification results and it becomes the inputs of the prediction experiment. In the second experiment, we use the features that we find in the first experiment as the ANFIS inputs and calculate the correlation, MSE, regression slop under different membership functions. In the third experiment, we combine the features of radar system with our features to perform classification and prediction again.

6.1. Classification with support vector machine

Support vector machine is one of the most widely used machine-learning algorithms for classification problems [22].

We group the subjects of level 1 and level 2 into the Bad group, and the subjects of level 3 belong to the Good group. There are 32 subjects in the Bad group and 28 subjects in the Good group. Those subjects who belong to the Good group are marked with blue triangles and those subjects who belong to the Bad group are marked with red circles.

S-value	1	2	3	4	5	6	7
Accuracy	0.55	0.55	0.55	0.56	0.56	**0.61**	0.58

Table 2. The SVM accuracy in different *S*-value with $\sigma'_{D(1,S)}$ and $\sigma'_{V(1,S)}$.

According to the above chapter, we divide all steps into *S* sections. In order to find the *S*-value, we perform different *S*-values in SVM classification. The inputs are $\sigma'_{D(1,S)}$ and $\sigma'_{V(1,S)}$ because these two parameters are affected by the *S*-value. **Table 2** lists the SVM accuracy in different *S*-values and the highest accuracy is **0.61** when *S* equals six. Therefore, the *S*-value is six in our experiment. In this article, the bold values in all the tables means the best result in the experiment.

Level	D_p'	V_p'
Level 1	47.72	117.4
Level 2	50.49	129.2
Level 3	**56.55**	**152.3**

Table 3. The D_p' and V_p' values.

http://dx.doi.org/10.5772/63917

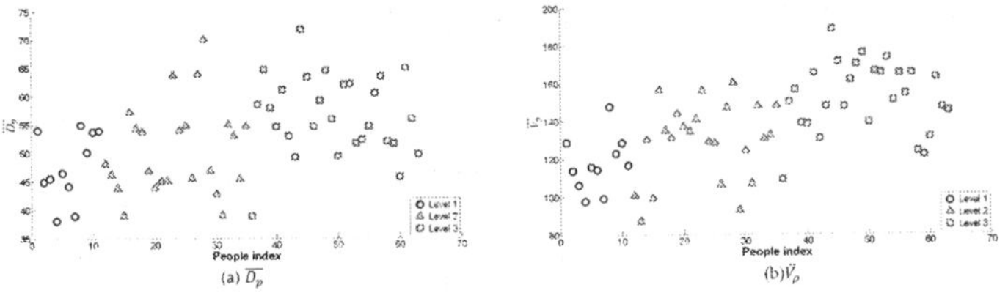

Figure 15. The D_p' and V_p' values of all subjects.

From **Table 3**, the subjects of level 3 have larger D_p' and V_p' than that of levels 2 and 1. The values of each level are the mean of D_p' and V_p' of the subjects who belong to the level. Therefore, the D_p' and V_p' become our choices of input features. **Figure 15** shows the D_p' and V_p' of all subjects.

Level	$\sigma_{D(1,6)}'$	$\sigma_{V(1,6)}'$
Level 1	3.16	11.2
Level 2	3.48	11.9
Level 3	**3.69**	**14.3**

Table 4. The $\sigma_{D(1,6)}'$ and $\sigma_{V(1,6)}'$ values in three levels.

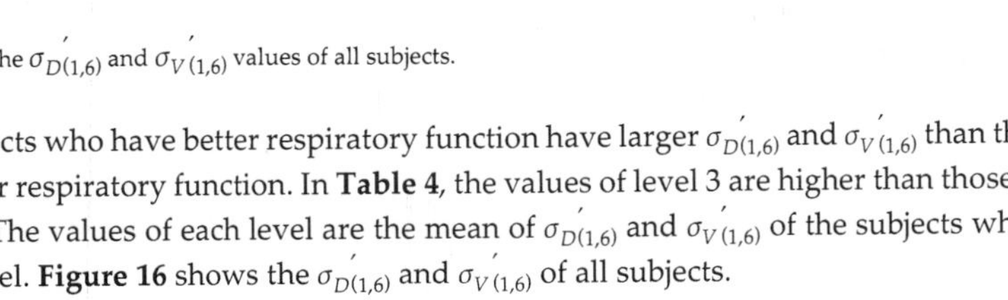

Figure 16. The $\sigma_{D(1,6)}'$ and $\sigma_{V(1,6)}'$ values of all subjects.

The subjects who have better respiratory function have larger $\sigma_{D(1,6)}'$ and $\sigma_{V(1,6)}'$ than those who have poor respiratory function. In **Table 4**, the values of level 3 are higher than those of levels 1 and 2. The values of each level are the mean of $\sigma_{D(1,6)}'$ and $\sigma_{V(1,6)}'$ of the subjects who belong to the level. **Figure 16** shows the $\sigma_{D(1,6)}'$ and $\sigma_{V(1,6)}'$ of all subjects.

In addition, people have lower variance after they exercise. **Figure 17** shows the pace distance of a person in three different conditions: normal walking, walking after a short run and walking after a long run. The variation of walking after a long run is smaller than others.

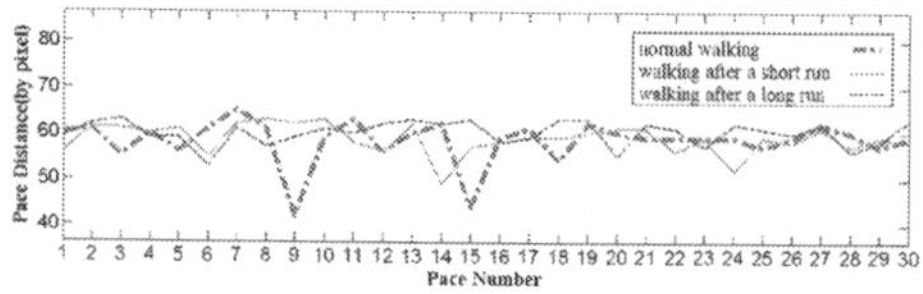

Figure 17. Pace distance of a person in three conditions.

The $\sigma'_{V(5,6)}$ is the mean of velocity variance of the hind of the test and $\sigma'_{V(1,2)}$ is the mean of velocity variance of the front of the test. To those subjects who have poor respiratory function, the $\sigma'_{V(5,6)}$ is smaller than $\sigma'_{V(1,2)}$ because they feel like walking after a long run in the hind of the test. On the other hand, for those subjects who have better respiratory function feel like normal walking in the hind of the test.

Therefore, the $\sigma'_{V(5,6)}/\sigma'_{V(1,2)}$ of those subjects who have poor respiratory function should be smaller than those subjects who have better respiratory function. From **Table 5**, the $\sigma'_{V(5,6)}/\sigma'_{V(1,2)}$ value of level 3 is greater than levels 1 and 2. The values of each level are the mean of $\sigma'_{V(5,6)}/\sigma'_{V(1,2)}$ of the subjects who belong to the level. **Figure 18** shows the $\sigma'_{V(5,6)}/\sigma'_{V(1,2)}$ of all subjects.

Level	$\frac{\sigma'_{V(5,6)}}{\sigma'_{V(1,2)}}$
Level 1	0.93
Level 2	0.96
Level 3	**1.04**

Table 5. The $\frac{\sigma'_{V(5,6)}}{\sigma'_{V(1,2)}}$ values in three levels.

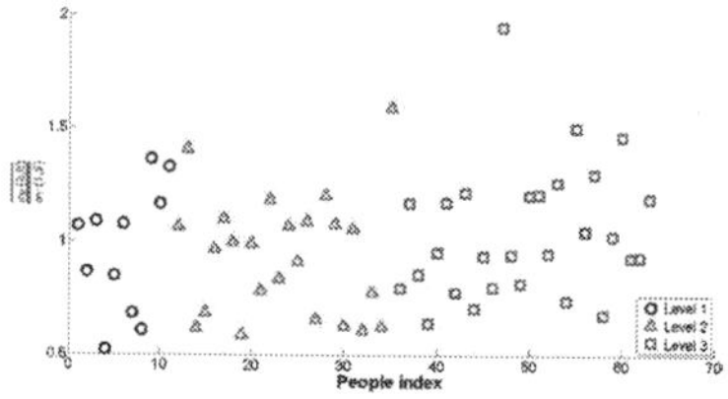

Figure 18. The $\frac{\sigma'_{V(5,6)}}{\sigma'_{V(1,2)}}$ of all subjects.

http://dx.doi.org/10.5772/63917

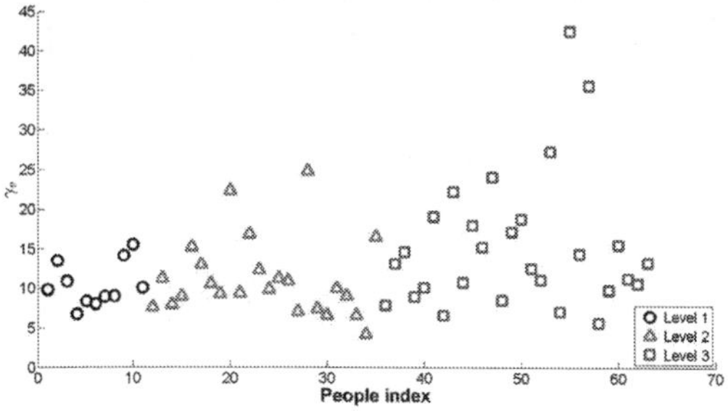

Figure 19. The γ_V of all subjects.

Input features	D_p', γ_V	V_p', γ_V
Accuracy	0.66	**0.75**

Table 6. The accuracy of SVM with different features.

The γ_V considers the mean of velocity variance value and the velocity variance ratio ($\gamma_V = \sigma'_{V(1,6)} * (\sigma'_{V(5,6)} / \sigma'_{V(1,2)})$), so γ_V become our choices of the input features. **Figure 19** shows γ_V values of all subjects. **Table 6** lists the SVM results with different features.

The accuracy of input features (V_p', γ_V) is better than the other input features (D_p', γ_V). Therefore, we use (V_p', γ_V) as the best inputs of the classification experiment. **Figures 20 and 21** show the SVM result with the inputs (V_p', γ_V) and (D_p', γ_V), respectively. In **Figure 20**, the subjects of the Good group have higher V_p' and γ_V than those who are in the Bad group.

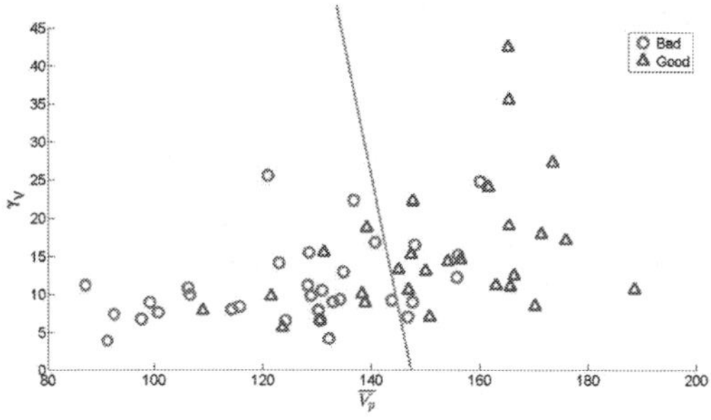

Figure 20. SVM classification result with V_p' and γ_V.

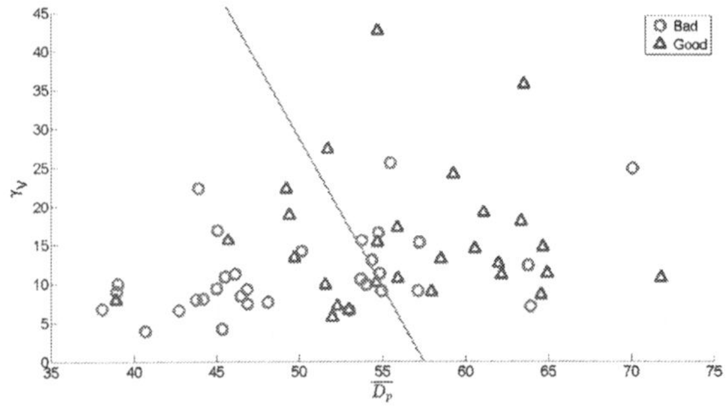

Figure 21. SVM classification result with $\acute{D}_p$ and γ_V.

6.2. Prediction with adaptive neural fuzzy inference systems

We utilize adaptive neural fuzzy inference system to help us predict the parameters from pulmonary spirometer. The ANFIS system comes from the toolbox of Matlab.

Because we only collect about 60 cases so far, it is not enough for ANFIS to perform prediction. In order to increase the training samples, we adopt **Leave-one-out cross-validation** method. **Leave-one-out cross-validation** is used in analysing small datasets. It uses one sample as the validation set and the remaining as the training set. Repeat on this way for all samples. We can use this method to solve the insufficient data problem.

In ANFIS, it is important to choose a correct membership function. In addition, we also need to choose the input sections. In the experiment, we use six different membership functions including trapmf, gbellmf, gaussmf, gauss2mf, pimf and dsigmf. **Figure 22** shows the membership functions we use in our prediction experiment. The inputs of the experiment are $\acute{V}_p$ and γ_V. In our results, we show the correlation, normalized Mean Square Error (MSE_N) and regression slope under different membership functions. The formula of MSE_N is shown in Eq. (20). The $Target_i$ are the measured values from pulmonary spirometer and the $Predict_i$ are the values come from ANFIS.

$$MSE_N = \frac{1}{n}\sum_{i=1}^{n}\left(\frac{Target_i - Predict_i}{Mean(Target)}\right)^2 \quad (20)$$

We try to predict three different parameters that come from the pulmonary spirometer: post $\frac{FEV1}{FVC}$, postFEV1 and postFVC. The 'post' name means the parameters after the 6-min brisk walking test. In the following part, for the convenience, we call post $\frac{FEV1}{FVC}$, postFEV1, postFVC as, $\frac{FEV1}{FVC}$ FEV1, FVC, respectively. $\frac{FEV1}{FVC}$ and FEV1 are used to access the respiratory function, so we choose these two parameters as our predicting targets.

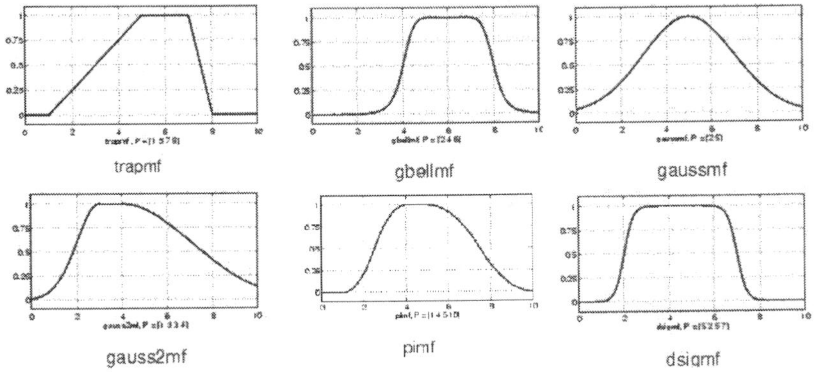

Figure 22. Different membership functions.

	Correlation	MSE_N	Regression slope
Trapmf	0.226	0.040	0.139
Gbellmf	0.203	0.041	0.127
Gaussmf	**0.251**	**0.040**	**0.167**
Gauss2mf	0.195	0.040	0.112
Pimf	0.215	0.039	0.125
Dsigmf	0.210	0.039	0.121

Table 7. The results of predicting $\frac{\text{FEV1}}{\text{FVC}}$.

FEV1-FVC is an index which is used to access the severity of airway obstruction. The lower value means that the airway obstructs severely. **Table 7** shows our prediction results and we use [2 2] as the input sections. The V_p' and γ_V are the experiment inputs. **Figure 23** shows the predicting results and regression slope under different membership functions.

FEV1 is also a parameter to access the respiratory function. The higher FEV1 value means that the subjects have better respiratory function. Consequently, we also predict the FEV1 value. **Table 8** shows the prediction results and we use [3 2] as using 3 nodes for the first input and 2 nodes for the second input in the ANFIS system. The features V_p' and γ_V are the experiment inputs. **Figure 24** shows the predicting results and regression slope in different membership functions.

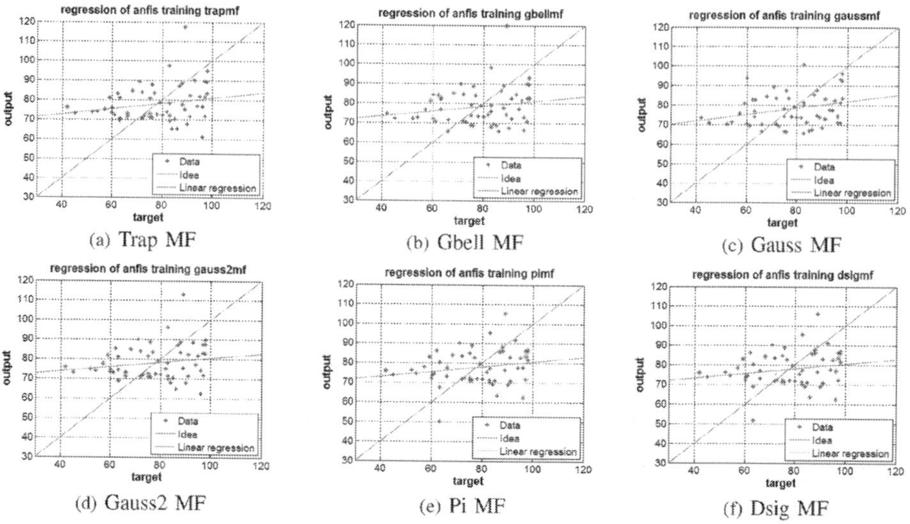

Figure 23. The results of predicting $\frac{\text{FEV1}}{\text{FVC}}$ with different membership functions: (a) Trap MF, (b) Gbell MF, (c) Gauss MF, (d) Gauss2 MF, (e) Pi MF and (f) Dsig MF.

	Correlation	**MSE$_N$**	**Regression slope**
Trapmf	**0.694**	**0.140**	**0.746**
Gbellmf	0.668	0.186	0.827
Gaussmf	0.640	0.193	0.774
Gauss2mf	0.386	1.968	1.264
Pimf	0.560	0.365	0.880
Dsigmf	0.352	2.692	1.321

Table 8. The result of predicting FEV1.

The correlation and regression slope of predicting $\frac{\text{FEV1}}{\text{FVC}}$ do not perform well under all membership functions. However, the correlation and regression slope of predicting FEV1 perform well under trapmf. From the two results above, we infer that our system is good at predicting the respiratory parameter from pulmonary spirometer (FEV1). However, it is hard to predict the computed value ($\frac{FEV1}{\text{FVC}}$). Consequently, we also predict the FVC value. **Table 9** shows the prediction results and we use [3 2] as using 3 nodes for the first input and 2 nodes for the second input in the ANFIS system. $\acute{V}_p$ and γ_V are the experiment inputs.

Figure 25 shows the predicting results and regression slope in different membership functions.

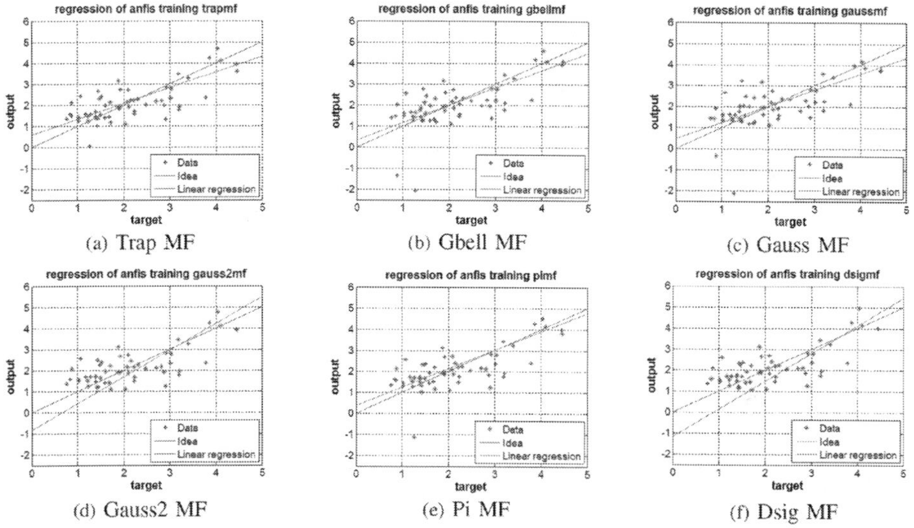

Figure 24. The results of predicting FEV1 with different membership functions: (a) Trap MF, (b) Gbell MF, (c) Gauss MF, (d) Gauss2 MF, (e) Pi MF and (f) Dsig MF.

	Correlation	**MSE_N**	**Regression slope**
Trapmf	0.660	0.079	0.621
Gbellmf	0.647	0.094	0.693
Gaussmf	**0.678**	**0.076**	**0.646**
Gauss2mf	0.432	0.249	0.654
Pimf	0.192	0.458	0.354
Dsigmf	0.156	1.151	0.469

Table 9. The result of predicting FVC.

Table 10 shows the best results of predicting $\frac{FEV1}{FVC}$, FEV1 and FVC. The correlations of FEV1 and FVC are all good and close and the regression slope of FEV1 is better than FVC. However, the correlation and regression slope of $\frac{FEV1}{FVC}$ do not perform well. The MSE_N of $\frac{FEV1}{FVC}$ is smaller than other two parameters. Our system is good at predicting the parameters of pulmonary spirometer but do not perform well in predicting the ratio. Nevertheless, the high correlations of FEV1 and FVC verify that there is a correlation between pulmonary spirometer and our gait analysis system.

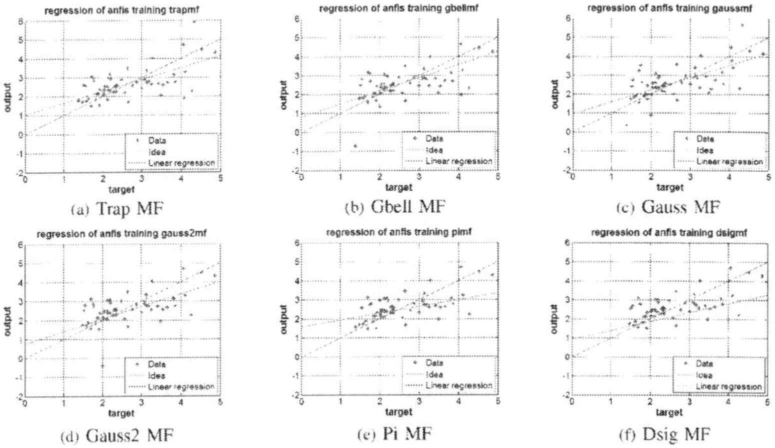

(a) Trap MF (b) Gbell MF (c) Gauss MF

(d) Gauss2 MF (e) Pi MF (f) Dsig MF

Figure 25. The results of predicting FVC with different membership functions: (a) Trap MF, (b) Gbell MF, (c) Gauss MF, (d) Gauss2 MF, (e) Pi MF and (f) Dsig MF.

Predicting target	Correlation	MSE_N	Regression slope
$\frac{FEV1}{FVC}$	0.251	**0.040**	0.167
FEV1	**0.694**	0.140	**0.746**
FVC	**0.678**	0.076	0.646

Table 10. The best result of predicting $\frac{FEV1}{FVC}$, FEV1 and FVC.

6.3. Cooperating with radar system

In this section, we combine the features of radar system with our features $\acute{V}_p$ and γ_V. The radar system is a tool using impulse to record the moving of the subjects' chest and analyse the features of respiration. It uses the ΔAmp and the $\Delta\beta$ratio to analyse respiration. These two features are listed in Eq. (21). The Amp is the respiratory intensity of the subject. The β_1 and β_2 are the inspiratory speed and expiratory speed, respectively. The names 'post' and 'pre' mean the parameters after a 6-min brisk walking and before a 6-min brisk walking, respectively. We use these two features with $\acute{V}_p$ and γ_V to perform SVM classification and ANFIS prediction experiments.

http://dx.doi.org/10.5772/63917

In the SVM experiment, there are still 32 subjects in the Bad group and 28 subjects in the Good group. Because we use ($\acute{V}_p$, γ_V, ΔAmp,$\Delta\beta$ratio) as the SVM inputs, we cannot draw a two-dimensional (2D) figure. The accuracy with ($\acute{V}_p$, γ_V, ΔAmp, $\Delta\beta$ratio) is **81.6%** and it is higher than the accuracy with ($\acute{V}_p$, γ_V) (75%).

$$\begin{cases} \Delta Amp = \dfrac{postAmp - preAmp}{preAmp} * 100 \\ \Delta\beta ratio = \dfrac{post\beta ratio - pre\beta ratio}{pre\beta ratio} * 100 \\ \beta ratio = \beta_1 / \beta_2 \end{cases} \tag{21}$$

Target (features)	Correlation	MSE_N	regression slope
$\frac{FEV1}{FVC}$, ($\acute{V}_p$, γ_V)	0.251	**0.040**	0.167
$\frac{FEV1}{FVC}$, (ΔAmp, $\Delta\beta$ratio)	**0.525**	0.071	**0.84**
$\frac{FEV1}{FVC}$, ($\acute{V}_p$, γ_V, ΔAmp, $\Delta\beta$ratio)	0.428	0.054	0.534
FEV1, ($\acute{V}_p$, γ_V)	**0.694**	**0.140**	0.746
FEV1, (ΔAmp,$\Delta\beta$ratio)	0.474	0.190	0.39
FEV1, ($\acute{V}_p$, γ_V, ΔAmp, $\Delta\beta$ratio)	0.675	0.020	**0.864**
FVC, ($\acute{V}_p$, γ_V)	**0.678**	**0.076**	0.646
FVC, (ΔAmp,$\Delta\beta$ratio)	0.129	0.217	0.13
FVC, ($\acute{V}_p$, γ_V, ΔAmp, $\Delta\beta$ratio)	0.517	0.190	**0.719**

Table 11. The best results of predicting $\frac{FEV1}{FVC}$, FEV1 and FVC.

In the ANFIS experiment, we also predict $\frac{FEV1}{FVC}$ and FEV1 parameters and use ($\acute{V}_p$, γ_V, ΔAmp, $\Delta\beta$ratio) as the ANFIS experiment inputs. The input sections we used is [3 3 3 2] that means the number of nodes used in the four inputs are 3, 3, 3, and 2 respectively. **Table 11** shows the results of predicting $\frac{FEV1}{FVC}$, FEV1 and FVC with ($\acute{V}_p$, γ_V), (ΔAmp, $\Delta\beta$ratio) and ($\acute{V}_p$, γ_V, ΔAmp, $\Delta\beta$ratio). We only list the best results among the six different membership functions. **Figure 26** shows the best results of predicting $\frac{FEV1}{FVC}$, FEV1 and FVC with ($\acute{V}_p$, γ_V), (ΔAmp, $\Delta\beta$ratio) and ($\acute{V}_p$, γ_V, ΔAmp, $\Delta\beta$ratio). In predicting $\frac{FEV1}{FVC}$, the correlation and regression slope improve strongly though the MSE_N value increases slightly by using ($\acute{V}_p$, γ_V, ΔAmp, $\Delta\beta$ratio). In predicting FEV1 and FVC, it does not improve the effects on correlation and regression slope by using ($\overline{V}_p$, γ_V, ΔAmp, $\Delta\beta$ratio). Therefore, the features of radar system cannot improve the results of predicting FEV1 and FVC.

Radar system improves our analysis results on both SVM classification and predicting the parameter *FEV1/FVC*. With radar system's help, there is a higher correlation and accuracy between the combined system and the pulmonary spirometer.

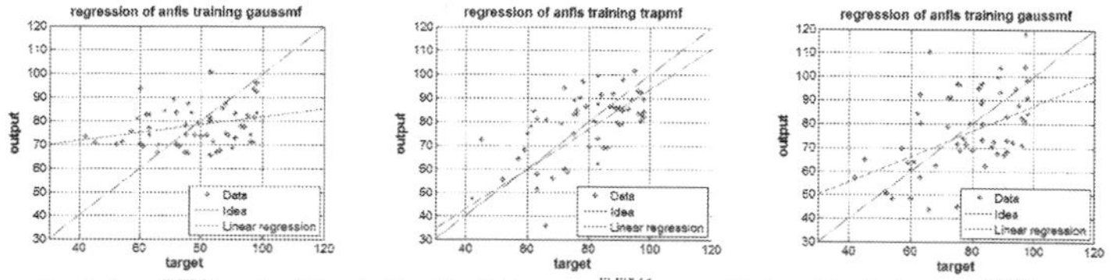

(a) Predicting $\frac{FEV1}{FVC}$ with ($\overline{V_p}$,γ_V) (b) Predicting $\frac{FEV1}{FVC}$ with (ΔAmp,$\Delta\beta$ $ratio$) (c) Predicting $\frac{FEV1}{FVC}$ with ($\overline{V_p}$,γ_V,ΔAmp,$\Delta\beta$ $ratio$)

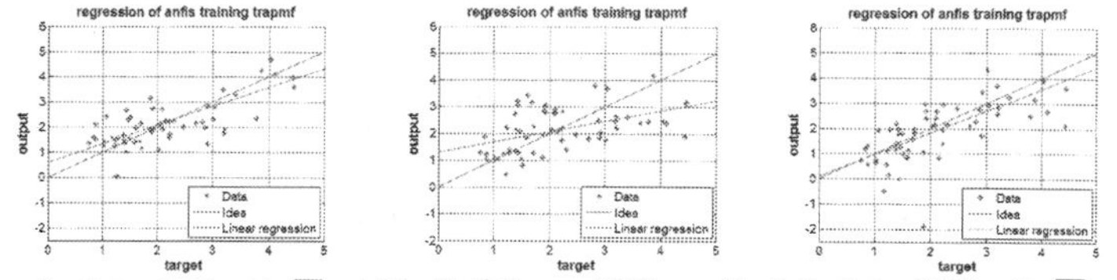

(d) Predicting FEV1 with ($\overline{V_p}$,γ_V) (e) Predicting FEV1 with (ΔAmp,$\Delta\beta$ $ratio$) (f) Predicting FEV1 with ($\overline{V_p}$, γ_V, ΔAmp, $\Delta\beta$ $ratio$)

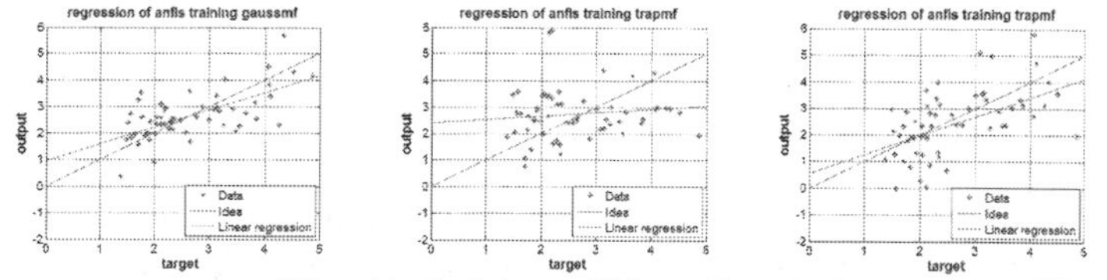

(g) Predicting FVC with ($\overline{V_p}$,γ_V) (h) Predicting FVC with (ΔAmp,$\Delta\beta$ $ratio$) (i) Predicting FVC with ($\overline{V_p}$, γ_V, ΔAmp, $\Delta\beta$ $ratio$)

Figure 26. (a) Predicting $\frac{\text{FEV1}}{\text{FVC}}$ with ($\overline{V}_p$, γ_V); (b)Predicting $\frac{\text{FEV1}}{\text{FVC}}$ with (ΔAmp, $\Delta\beta$ratio); (c) Predicting $\frac{\text{FEV1}}{\text{FVC}}$ with ($\overline{V}_p$, γ_V, ΔAmp, $\Delta\beta$ratio); (d) Predicting FEV1 with ($\overline{V}_p$, γ_V); (e) Predicting FEV1 with (ΔAmp, $\Delta\beta$ratio); (f) Predicting FEV1 with ($\overline{V}_p$, γ_V, ΔAmp, $\Delta\beta$ratio); (g) Predicting FVC with ($\overline{V}_p$, γ_V); (h) Predicting FVC with (ΔAmp, $\Delta\beta$ratio); (i) Predicting FVC with ($\overline{V}_p$, γ_V, ΔAmp, $\Delta\beta$ratio).

7. Conclusion

We propose a vision sensor-based gait analysis method without wearing any sensor on human body. In our approach, the proposed gait features analyse the subjects' respiratory function. We also perform a clinical experiment on COAD patients and normal people with our vision sensor-based gait analysis method. With the extracted features, $\overline{V}_p$ and γ_V, the classification result is close to the classification by the parameters of pulmonary spirometer. The SVM accuracy is **75%**. In ANFIS experiment, the correlations of ANFIS prediction on FEV1 and FVC achieve **0.694** and **0.678**.

In addition, by combining the features of radar system (ΔAmp and $\Delta\beta$ratio) with our features (V_p' and γ_V), the SVM accuracy and predicting on ratio ($\frac{\text{FEV1}}{\text{FVC}}$) both improve strongly. The SVM accuracy goes to **81** from 75% and the correlation of ANFIS on predicting $\frac{\text{FEV1}}{\text{FVC}}$ goes to **0.428** from 0.25. From the experiment above, we verify that there exists a correlation between the pulmonary spirometer and gait analysis system.

Author details

Yu Sheng Chan[1], Wen Te Liu[2] and Ching Te Chiu[1*]

*Address all correspondence to: ctchiu@cs.nthu.edu.tw

1 Department of Computer Science, National Tsing Hua University, Hsinchu, Taiwan, ROC

2 School of Respiratory Therapy, Taipei Medical University, Taipei, Taiwan, ROC

References

[1] Yunfeng Wu, Sridhar Krishnan. Statistical Analysis of Gait Rhythm in Patients with Parkinson's Disease. IEEE Transactions on Neural Systems and Rehabilitation Engineering. 2010;18(2):150–158. DOI: 10.1109/TNSRE.2009.2033062

[2] Ming-Feng Wu, Chao-Ling Chen, Chih-Yu Wen, and Jeng-Yuan Hsu. Design of Pervasive Rehabilitation Monitoring for COPD. IEEE Sensors Journal. 2013;13(11): 4413–4422.

[3] Howell AM, Kobayashi T, Hayes HA, Foreman KB, and Bamberg SJM. Kinetic Gait Analysis Using a Low-Cost Insole. IEEE Transactions on Biomedical Engineering. 2013;60(12):3284–3290.

[4] Fabbri LM and Hurd SS. Global Strategy for the Diagnosis, Management and Prevention of COPD: 2003 update. European Respiratory Journal. 2003;22(1):1–2.

[5] World Health Organization. The World Health Report 2000. 2000.

[6] Murray CJ and Lopez AD. Mortality by Cause for Eight Regions of the World: Global Burden of Disease Study. Lancet. 1997;349(9061):1269–1276.

[7] ATS/ERS Task Force on Pulmonary Rehabilitation. An Official European Respiratory Society Statement on Physical Activity in COPD. European Respiratory Journal. 2014;44(6):1521–1537.

[8] Sullivan SD, Ramsey SD, and Lee TA. The Economic Burden of COPD. Chest. 2000;117:5S–9S.

[9] Vorrink SN, Kort HS, Troosters T, and Lammers JW. Level of Daily Physical Activity in Individuals with COPD Compared with Healthy Controls. Respiratory Research. 2011;12-33. DOI: 10.1186/1465-9921-12-33.

[10] Gimeno-Santos E., et al. Determinants and Outcomes of Physical Activity in Patients with COPD: A Systematic Review. Thorax. 2014;69(8):731–739.

[11] Shu-Yi Liao, Roberto Benzo, Andrew L. Ries, and Xavier Soler. Physical Activity Monitoring in Patients with COPD. Journal of the COPD Foundation. 2014;1(2):155–165.

[12] Mario Merone, Leonardo Onofri, Paolo Soda, Claudio Pedone, Raffaele Antonelli Incalziand, and Giulio Iannello. Early Experiences in COPD Exacerbation Detection. In: IEEE International Symposium on Computer-Based Medical Systems, pp. 406–410.

[13] Bellos CC, Papadopoulos A, Rosso R, and Fotiadis DI. Identification of COPD Patients' Health Status Using an Intelligent System in the CHRONIOUS Wearable Platform. IEEE Journal of Biomedical and Health Informatics. 2014;18(3):731–738.

[14] Wen L, Qian J, Hu X, Shen L, Wu X, and Yu C. Gait Measurement and Quantitative Analysis in Patients with Parkinson's Disease for Rehabilitation Assessment. In: Robotics and Biomimetics (ROBIO), IEEE International Conference on. IEEE; 2013. Shenzhen, China. pp. 286–291.

[15] Chung PC, Hsu YL, Wang CY, Lin CW, Wang JS, and Pai MC. Gait Analysis for Patients with Alzheimer'S Disease using a Triaxial Accelerometer. In: IEEE International Symposium on Circuits and Systems. IEEE; 2012. Seoul, Korea. pp. 1323–1326.

[16] Susu Jiang, Bofeng Zhang, Guobing Zou, and Daming Wei. The Possibility of Normal Gait Analysis Based on a Smart Phone for Healthcare. In: Green Computing and Communications, IEEE and Internet of Things, IEEE International Conference on and IEEE Cyber, Physical and Social Computing. IEEE; 2013. Beijing, China. pp. 2235–2240.

[17] Jamie Schiel and Richard Green. Adaptive Human Silhouette Extraction with Chromatic Distortion and Contour Tracking. In: International Conference on Image and Vision Computing New Zealand (IVCNZ 2013); Wellington. IEEE; 2013. pp. 288–292.

[18] Vladimir N. Vapnik. The Nature of Statistical Learning Theory. New York, NY: Springer; 1995. DOI: 10.1007/978-1-4757-3264-1

[19] Jang JSR. ANFIS: Adaptive-Network-Based Fuzzy Inference System. IEEE Transactions on Systems, Man and Cybernetics. 1993;23(3):665–685.

[20] Al-Hmouz A, Shen J, Al-Hmouz R, and Yan J. Modeling and Simulation of an Adaptive Neuro-Fuzzy Inference System (ANFIS) for Mobile Learning. IEEE Transactions on Learning Technologies. 2012;5(3):226–237.

[21] Cleveland Clinic Foundation. 2003. Available from: http://www.clevelandclinicmeded.com/medicalpubs/diseasemanagement/pulmonary/pulmonary-function-testing/ March 2015.

[22] Jie Xu, Yuan Yan Tang, Bin Zou, Zongben Xu, Luoqing Li, and Yang Lu. The Generalization Ability of Online SVM Classification Based on Markov Sampling. IEEE Transactions on Neural Networks and Learning Systems. 2015;26(3):628–639.

Chapter 2

Position-Free Vital Sign Monitoring: Measurements and Processing

Dany Obeid, Sarah Samad, Sawsan Sadek,
Gheorghe Zaharia and Ghaïs El Zein

Additional information is available at the end of the chapter

http://dx.doi.org/10.5772/63915

Abstract

As traditional electrodes are perturbing for patients in critical cases such as for burn victims or newborn infants, and even to detect life sign under rubble, a contactless monitoring system for the life signs is a necessity. The aim of this chapter is to present a complete process used in detecting cardiopulmonary activities. This includes a microwave Doppler radar system that detects the body wall motion and signal processing techniques in order to extract the heartbeat rate. Measurements are performed at different positions simultaneously with a PC-based electrocardiogram (ECG). For a distance of 1 m between the subject and the antennas, measurements are performed for breathing subject at four positions: front, back, left, and right. Discrete wavelet transform is used to extract the heartbeat signal from the cardiopulmonary signal. The proposed system and signal processing techniques show high accuracy in detecting the cardiopulmonary signals and extracting the heartbeat rate.

Keywords: Doppler radar, cardiopulmonary signals, wavelet transforms, electrocardiogram, contactless monitoring

1. Introduction

Traditional electrocardiogram (ECG) with affixed electrodes could be perturbing for patients with conditions such as burn victims or newly born infants, or when long duration monitoring is needed. In addition, a monitoring system detecting life signs under rubble or snow is helpful, especially after earthquakes where the detection of life signs over long distance is needed. The utility of microwave Doppler radar used in the detection of life signs has recently increased. Hence, a touchless cardiopulmonary monitoring is needed for such applications and could be a prominent tool in home health care applications.

When a radio wave is transmitted toward a person, it will be reflected off his/her body. If the reflection occurs off the chest of a motionless person, the reflected signal has a phase modulation due to the Doppler effect due to the movement of the chest, which is caused mainly by heartbeats and breathing. On the other hand, when the breath is held, the reflected signal will depend on the chest displacement due to heartbeat alone.

The aim of this chapter is to present a tunable system in terms of power and frequency that shows the capability to detect chest displacement due to heart beat at different operational frequencies and for different transmitted powers. This allows specifying the appropriate operational frequency for the minimum transmitted power. In addition, simultaneously with a PC-based ECG, measurements are performed at the four different sides (front, back, left, and right) of the person under test (PUT). Wavelet transforms are used in order to separate the heartbeat signal from the cardiopulmonary signal and to extract the heartbeat rate (HR).

The rest of this chapter is organized as follows: Section 2 provides background information about chest-wall displacement. Section 3 describes related work for the system design and the signal processing technique. Section 4 presents the proposed system and some preliminary results obtained at different operational frequencies. Section 5 shows the cardiopulmonary signals obtained at different sides from the subject and presents the results upon applying the proposed signal processing technique. Section 6 concludes the work.

2. Chest displacement due to breathing and heart beating

Using microwave Doppler radar, several techniques were established in order to sense the cardiopulmonary activity. When a microwave signal is transmitted to a person's chest, the power of the reflected signal when it occurs at the air/skin interface is higher than the power of the signal reflected from internal organs. Then, the signal reflected off the person's chest contains information about the chest displacement due to cardiopulmonary activity including breathing and heartbeat. Based on these displacements, the heartbeat rate and the respiration rate can be extracted. However, these motions are not the same for all people. This section describes the mechanism of the chest displacement due to both heartbeat and respiration, as well as some experiments measuring the displacement amount.

2.1. Surface motion due to the cardiac cycle

When the heart beats, it pushes blood through the lungs and to tissues throughout the whole body. A pressure is generated when the heart contracts in order to drive the flow of the blood. While contracting, the heart hits the cavity of the chest creating a significant displacement at the surface of the skin. As the left ventricle carries out blood to all parts of the body, the contraction and relaxation of the left ventricle cause a larger chest motion than other heart actions in healthy subjects. During isovolumetric contraction, the heart normally undergoes a partial rotation in a counterclockwise direction, causing the lower front part of the left ventricle to strike the front of the chest wall [1]. Also, the left ventricle shortens while contracting, shaping the heart to be more spherical, increasing its diameter, and further adding to the impulse on the chest wall [2]. The peak outward motion of the left ventricular impulse occurs either simultaneously with or just after the opening of the aortic valve. Then the left ventricular apex moves inward [1]. The left ventricular motion causes the chest to pulse outward briefly and the adjacent chest retracts during ventricular ejection [3]. This impulse occurs at the lowest point on the chest where the cardiac beat can be seen, and it is normally above the anatomical apex [4]. Some studies found a second outward movement at the apex: the pre-ejection beat [5]. Many techniques for quantitatively measuring the gross displacement of the chest wall have been applied, including the impulse cardiogram [1], a single-point laser displacement system [6], structured lights and the Moiré effect [7], laser speckle interferometry [8], a capacitance transducer [9], a magnetic displacement sensor [10], and a phonocardiographic microphone [11]. The values of the skin motion due to heartbeat vary between individuals due to physiological difference, age differences, and body shape differences. Since the amount and the speed of the motion of the heart within the chest changes with age, it is expected that the amount of the chest motion due to the heartbeat changes with age. Recent studies found that the expected movement of the mitral valve ahead the heart's long axis is about 1.49 cm at age of 20 with an expected velocity of 7.48 cm/s, and 1.22 cm at age of 84 with an expected velocity of 7.48 cm/s [12]. Another study showed that the displacement of the septum decreases with age, but the displacement of the left lateral wall and the posterior wall of the heart remains constant between ages 49 and 73 [13]. The absolute diastolic displacement of annular sites among children increases significantly with increasing body weight (which is expected since the size of the heart and thorax is increasing), but the percent displacement was inversely proportional to body weight [14]. Although there is no significant change in left ventricular ejection volume with age, the arterial pressure wave varies greatly with age: arterial wall thickness increases, arterial diameter increases, and arterial distensibility decreases [15]. The arterial wall rigidity is expressed as [16]

$$\alpha = 0.421 + 0.0602 \times \text{age} \tag{1}$$

This indicates that pulses will be smaller and more difficult to measure in older subjects. At a pulse, subjects aged between 60 and 70 years have 50% lower variation in the cross-sectional area of the artery than subjects aged between 20 and 30 years [16]. When the variation in the arteries' diameter decreases, the amplitude of the surface skin displacement will decrease. This results in decreasing the signal-to-noise ratio when measuring the pulse using a Doppler radar. In average, the peak-to-peak chest motion in adults due to the heartbeat is about 0.5 mm.

2.2. Surface motion due to respiration

The chest surface motion due to breathing is the combination of the abdominal and rib cage movements. A linear correlation exists between cross-sectional area of the thorax, displacement of the diaphragm, displacement of the rib cage, and lung volume [17]. At the third rib, the angle of the pump handle changes between 20° and 30°, and the rib radius changes between 10.6 and 10.8 cm. At the seventh rib, the angle of the pump handle changes between 30° and 37°, and the rib radius changes between 1.37 and 1.42 cm [18]. Comparing the chest motion in the front/back, left/right, and up/down directions shows that the largest motions correspond to the sternum and the navel. Sternum moves forward 4.3 mm with inspiration, and the navel which moves forward 4.03 mm with inspiration [19]. The relation between tidal volume and abdominal wall linear displacement is measured using a laser displacement measuring device [20]. An expansion of the abdomen is observed: 4 mm for 400 ml inspiration and 11 mm for 1100 ml inspiration. Also, during spontaneous breathing, an abdominal displacement of 12 mm is observed.

3. Related work

This section presents the related work in both system design and signal processing techniques.

3.1. System design

Since the 1970s [21], microwave Doppler radar has been used in sensing physiological movement. The original work was done with heavy, bulky, and expensive components. However, it was useful for research improvements. During the 1980s, heart and respiration signals were obtained using 10.5 GHz frequency signal. Using a horn antenna placed few centimeters from the subject, the system shows capability to detect cardiopulmonary signals using a 10-mW transmitted power [22]. In 1990, systems operating at 2 and 10 GHz were tested in detecting life signs in victims under clutter. The radiated power varied between 10 and 20 mW [23]. In 1997, heartbeat and respiration signals were detected at a distance of 10 m using 24 GHz frequency system with 30 mW output power and 40 dB antenna gain [24]. In the year 2000, systems operating at 450 and 1150 MHz, with a radiated power around 300 mW, were used to detect life sings in victims under rubble [25]. In 2001, a 1.2-GHz, 70-mW quadrature superheterodyne system was used to detect breathing of a subject under 1.5-m rubble [26]. Operating at 1.6 and 2.4 GHz, direct-conversion Doppler radars have been integrated in 0.25 μm complementary metal-oxide semiconductor (CMOS) and BiCMOS technologies. The output power was estimated to be 6.5 dBm [27]. Heart and respiration activities were detected using a modified Wireless Local Area Network (LAN) Personal Computer Memory Card International Association (PCMCIA) card and a module combining the transmitted and reflected signals [28,29]. The operational power of the system was 35 mW and the distance from the subject was 40 cm. Other systems operating in the Ka-band were described in Ref. [30] using a low-power double-sideband transmission signal. For a distance of 2 m from the subject, and for 16 and 12.5 μW, respectively, the systems showed an accuracy of 80% in detecting the heartbeat rate. Recently in 2006, some measurements were performed in order to detect multiple heartbeats signals [31]. Operating at 2.4 GHz and 1 mW power, the system was able to determine the number of persons in a room. With the same characteristics, another system using single and multiple antennas systems showed the possibility of separating two respiration signals [32]. In 2007, a new study showed the possibility of detecting the presence of a person through a wall using ultra-wideband (UWB) radar [33]. Lately in 2008, some experiments are preformed to detect life signs using a 4–7-GHz band with 1 mW power and around 7 dB antenna gain. This system uses the complex signal demodulation (CSD) and the arctangent demodulation in order to cancel random body movements [34]. In 2009, a system operating at 10 GHz showed the ability to detect the heart and the respiration activity of a person behind a wall.

Recently, a system having two Vivaldi antennas, a Mini-Circuits ZHL-42 power amplifier for the transmission, and a Hittite HMC753 low-noise amplifier in the receiver is proposed [35]. The receiver is composed of a down-converter of a 20-MHz IF band, a mixer, an Agilent signal generator and a band-pass filter, and the received signal is sent to the analog-to-digital converter (ADC). A 60-MHz sampling clock provided by an external clock and synchronized with the field-programmable gate array (FPGA) reference clock for the signal digitization. Then, the sampled data are sent to the FPGA for digital down conversion. Another system composed of two antennas, an oscillator that provides both the receiver's local oscillator and the transmitted signal, and a mixer is presented in [36]. I/Q channel demodulation with calibration method is added to alleviate the null point problem and acquire an accurate phase demodulation result with high linearity. Another system presented in [37] is based on multiple transceivers, and antennas with polarization and frequency multiplexing are used to detect signals from different body orientations.

3.2. Signal processing techniques

In Doppler cardiopulmonary monitoring, the heartbeat and the respiration signals are laid together. Hence, a processing technique is needed in order to determine the characteristics of each signal. The signal processing part includes the separation of the cardiopulmonary signals and the extraction of the heartbeat rate. The amplitude of the respiration signal is much greater than the amplitude of the heartbeat signal. Therefore, the respiration rate can be determined without filtering. On the other hand, determining the heartbeat rate needs a processing technique. At rest, the heartbeat rate varies between 50 and 90 beats per minute [38]; this corresponds to a frequency between 0.83 and 1.5 Hz, respectively. On the other hand, the resting respiration rate varies between 9 and 24 breaths per minute [38]; this corresponds to a frequency between 0.15 and 0.4 Hz. Due to the difference of the frequencies that correspond to the heartbeat and the respiration rates, the average heartbeat rate could be determined upon extracting the frequency components of the cardiopulmonary signals. This allows determining the average heartbeat rate over a specific window of time. On the other hand, determining the heartbeats variation over time needs a peak-finding technique.

Several techniques were used in processing the cardiopulmonary signals. This processing includes separating the heartbeat signal from the respiration signal and then finding the heartbeat rate. Some measurements were performed while holding the breath. This eliminates the isolation process of the heartbeat signal, but a filtering approach is still needed in order to remove noise and distorting signals.

First measurements for heartbeat and respiration were performed separately. Holding the breath allows detecting the heartbeat signal [21]. Another study shows the possibility of measuring the heartbeat and respiration activities successively where a low-pass filter with 4

Hz cutoff frequency was used to remove unwanted frequencies [23]. In 2000, heartbeat and respiration signals were measured simultaneously. The output signal is fed through a band-pass filter (BPF) with passing band between 0.1 and 4 Hz. The heartbeat and respiration rates are obtained by applying fast Fourier transform (FFT) to the original signal. The dominant peak in the frequency domain was taken as the breathing frequency, and the second dominant peak was taken as the heartbeat frequency [25]. In 2002, separated measurements for heartbeat and respiration were performed. The respiration signal was filtered with a BPF (0.03–0.3 Hz), and the heart signal was filtered with a BPF (1–3 Hz) [27]. Another work tended to detect the heartbeat signal using a 12-dB/octave high-pass filtering at 0.03 Hz in order to remove DC offset, and a 12-dB/octave low-pass filtering at 3 Hz was used to avoid aliasing error. The heart signal was further isolated with an additional 12 dB/octave HPF at 1 Hz [39]. Also in 2002, measurements for breathing persons were performed. The respiration signal was isolated using a fourth-order low-pass Butterworth filter with cutoff frequency at 0.7 Hz. The heartbeat signal was isolated using a fourth-order band-pass Butterworth filter with cutoff frequencies at 1 and 3 Hz. The rate determination is based on the use of auto-correlation. A spatial zero-forcing filter is applied so that the DC is removed from the measured received signal [40]. In 2003, a wireless LAN PC card was used. A low-pass resistor-capacitor (RC) filter having a cutoff frequency 100 Hz is used to filter the baseband output of the receiver. This helps denoising the signal as well as avoiding aliasing error. The filtered signal is then converted to digital in order to be processed in a notebook PC. The prefiltered, digitized signal was filtered further in the digital domain to separate the heart and breathing signals. The heart signal was isolated using a 0.75–5 Hz band-pass filter for 10 s interval. Based on the periodicity of the autocorrelation function, the heartbeat rate was estimated [28]. In 2006, a system using a signal processing part similar to some previous work is stated. The heartbeat signal was first separated from the respiration signal by a Butterworth BPF with passband from 0.7 to 3 Hz. The filtered signal was then windowed and auto-correlated. Then, FFT was applied to the auto-correlated signal to obtain the heartbeat rate [41].

Recently, other processing techniques are used for cardiorespiratory separation. In [35], FPGAs are used to process either time- or frequency-domain signals in human sensing radar applications. It is applied for continuous wavelet (CW) and UWB radars. In CW Doppler radar, a novel superheterodyne receiver is used to suppress low-frequency noise and includes a digital down-converter module implemented in an FPGA. In [36], compact quadrature Doppler radar sensor is used: Continuous wavelet filter and ensemble empirical mode decomposition (EEMD) based algorithms are applied for cardiorespiratory signal to separate the cardiac and respiratory signals. The accurate beat-to-beat interval can be acquired in time domain for heart rate variability (HRV) analysis. A curvelet transform is applied in [42] in order to remove the direct coupling wave and background clutters. Life signals are denoised using a singular value decomposition. Both the FFT and the Hilbert-Huang transform are applied in order to separate and extract the frequencies of the human vital sign as well as the characteristics of micro-Doppler shift for an UWB radar. Least mean square adaptive harmonic cancellation algorithm is proposed in [43] to separate the breathing and heartbeat signal from biological Doppler radar. The respiration signal is used as a model reference input while the radar signal due to body motion is considered as the original input of the model. A model is designed and validated experimentally with commercial motion detector [44]. A low-pass filter with 0.7 Hz cutoff frequency is used to extract the respiration signal, while a band-pass filter between 0.9 and 2.5 Hz is used to extract the heartbeat signal. In [37], complex technique is discussed; a complex signal demodulation technique is proposed to eliminate the null detection point problem in non-contact vital sign detection. This technique is robust against DC offset in a direct conversion system. Hence, a random body movement cancellation technique is developed to cancel out strong noise caused by random body movement in non-contact vital sign monitoring. The complex signal is software reconstructed in real time by $S(t) = I(t) + \mathrm{j}\,Q(t)$. System setup of random body movement cancellation technique is designed of two transceivers, one in front of and the other behind the human body, which are transmitting and receiving signals with different polarization and wavelength. The two complex signals are multiplied. This multiplication corresponds to convolution and frequency shift in frequency domain, thus canceling the Doppler frequency drift and only keeping the periodic Doppler phase effects. In [45], fast acquisition of HR is proposed, the length of the time window is less than 5 s and the accuracy is significantly degraded due to insufficient spectrum resolution. In [37], CSD is used for vital sign detection. A time-window-variation technique is developed for fast acquisition of HR from short-period time windows and measuring HR variation using CSD. The proposed method has also proved to be able to measure HR variation using CSD.

3.3. Discussion

Systems used in these works lack determining the most appropriate parameters for these applications. These parameters are the operational frequency, the radiated power, and the optimal signal processing technique. The proposed system shows the ability of tuning both the operational frequency and transmitted power. Hence, it is able to determine better emitted frequency with less power that detects heartbeat accurately. On the other hand, most of the processing techniques tend to extract an average heartbeat rate of the subject. This does not give information about the variation of the heartbeat rate and requires a long-duration window which makes the real-time processing not possible. The proposed signal processing technique shows the ability to detect the variation of the heart activity in time.

4. Proposed system: design and preliminary results

The most important factors in a Doppler radar using a CW signal are the operational frequency and the radiated power. The former governs the penetration of clothing, the reflection at the air/skin interface, and the signal-to-noise ratio. The latter must be taken into consideration as both patients and medical staff are exposed to the radiations. Our proposed system shows the ability of tuning the operational frequency, as well as the transmitted power. The choice of the operational frequency and the radiated power are subject to international standards.

This section introduces the limitations in terms of power and frequency, presents the proposed system, and shows the preliminary results obtained at different operational frequencies: 2.4, 5.8, 10, 16, and 60 GHz.

4.1. Frequency and power limitations

The touchless cardiopulmonary detection is based on a reflected signal off the person's chest. Hence, the most important feature is that the RF signal penetrates clothing with minimal reflection and has a higher reflection at the air/skin interface. These properties depend on the operational frequency. For most microwave frequencies, the reflection at the air/skin interface is high. However, the signal reflects at the air/skin interface less as the frequency decreases, and it has more significant reflections from clothing or bedding as the frequency increases. As shown in Eq. (3), the amount of the phase variation is indirectly proportional to wavelength of the carrier. Hence, the signal-to-noise ratio is directly proportional to the operational frequency ($f = c/\lambda$). The higher the frequency, the shorter the wavelength, and then the greater the phase variation. For a constant phase noise at different frequencies, increasing the frequency increases the signal-to-noise ratio. Another feature related to the operational frequency is the antenna size. As the frequency increases, the same antenna gain can be obtained with a physically smaller antenna. The maximum directivity that can be obtained from an antenna with aperture area A is

$$D_{\max} = \frac{4\pi A}{\lambda^2} \tag{2}$$

where λ is the wavelength of the transmitted signal. When λ decreases, the area decreases for constant directivity. All these factors should be taken into consideration in order to specify the operational frequency.

As the Doppler radar results in a transmitted power, it is necessary to decrease the transmitted power in order to decrease the radiated energy the patient is exposed to during measurements. Various organizations and countries have developed exposure standards for radio frequency energy. These standards recommend safe levels of exposure for both the general public and for workers. Since 1985, In the USA, the Federal Communications Commission (FCC) has accepted and used approved safety guidelines for the exposure of the RF environmental. Several Federal health organizations such as the US Food and Drug Administration (FDA), the Environmental Protection Agency (EPA), the Occupational Safety and Health Administration (OSHA), and the National Institute for Occupational Safety and Health (NIOSH) have also been concerned in issues related to monitoring and RF exposure [46]. The FCC guidelines for human exposure to RF electromagnetic fields were derived from the recommendations of two expert organizations: the National Council on Radiation Protection and Measurements (NCRP) and the Institute of Electrical and Electronics Engineers (IEEE). Both the NCRP exposure criteria and the IEEE standard were developed by expert scientists and engineers after extensive reviews of the scientific literature related to RF biological effects.

The exposure guidelines are based on thresholds for known adverse effects, and they incorporate prudent margins of safety. In adopting the most recent RF exposure guidelines, the FCC consulted with the EPA, FDA, OSHA, and NIOSH and obtained their support for the guidelines that the FCC is using. The International Commission on Non-Ionizing Radiation Protection (ICNIRP) has developed exposure guidelines that are applicable in many countries including Europe. The safety limits stated by the ICNIRP limits are similar to those stated by the NCRP and IEEE, with few exceptions. For example, ICNIRP recommend different exposure thresholds in both the lower and the upper frequency ranges and for localized exposure due to some devices as cellular phones. The NCRP, IEEE, and ICNIRP exposure guidelines identify the same threshold level at which harmful biological effects may occur, and the values for maximum permissible exposure (MPE) recommended for electric and magnetic field strength and power density are based on this level. The American National Standards Institute (ANSI) standard was developed over a period of several years by scientists and engineers with considerable experience and knowledge in the area of RF biological effects and related issues. The recommendations were based on a determination that the threshold of hazardous biological effects was approximately 4 W/kg [47]. The Watts per kilogram unit is an expression for the rate of energy absorption in the body given in terms of the specific absorption rate (SAR). A safety factor of 10 was then incorporated to arrive at the final recommended protection guidelines. In other words, the protection guides can be correlated with an SAR threshold of about 0.4 W/kg [29]. In addition, the guidelines stated by NCRP, IEEE, and ICNIRP for the maximum permissible exposure depend on the transmitting frequencies. This is caused by the fact that the human body absorption of RF energy varies accordingly with the RF signal frequency. The highest RH energy absorbed by the human body lies in the frequency range 30–300 MHz; thus, the most restrictive limits are applied. Other exposure limits are stated for devices that expose only part of the body such as mobile phones [48].

4.2. System operating at different frequencies

The proposed system consists of using only a vector network analyzer (VNA) and two antennas. This accommodates a quick and simple installation process. **Figure 1** represents the proposed system. Beside simplicity, many benefits are offered upon the use of a VNA. This section describes the characteristics of the proposed system including VNA and antennas and shows the benefits of this system.

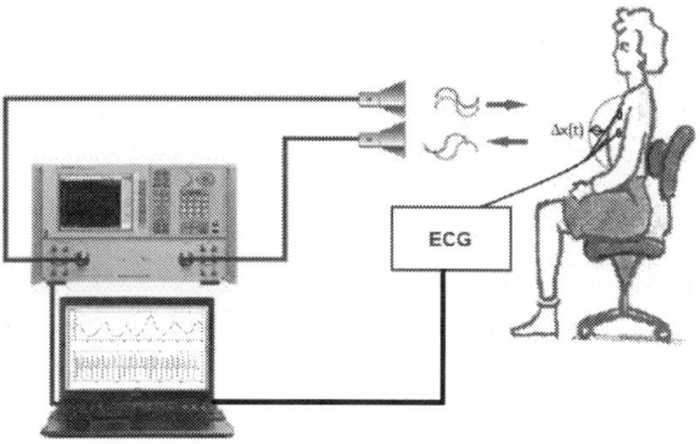

Figure 1. Proposed system design: microwave system and ECG.

In order to provide a comparative study in terms of operational frequency, several frequencies were tested using the proposed system. The operational frequencies should cover as much radar band designations as possible, as well as industrial-scientific-medical (ISM) bands. The operational frequencies chosen in this work are 2.4 GHz (ISM S-band), 5.8 GHz (ISM C-band), 10 GHz (X-band), 16 GHz (Ku-band), and 60 GHz (ISM V-band). Choosing these frequencies allows providing a comparative approach for different frequencies. In addition, the choice of the frequencies covers different bands of frequencies, specifically, S, C, X, Ku, and V bands. Also, the chosen frequencies include within some ISM bands, specifically 2.4, 5.8, and 60 GHz. Besides, these frequencies are selected taking into account the operational limits of the VNA and the antennas.

4.3. Vector network analyzer

VNA is one of the most used systems for microwave measurements and RF applications. It allows verifying the RF performance of microwave devices as well as their characterization in terms of network scattering parameters or *S* parameters in both magnitude and phase. The utilized vector network analyzer is an HP N5230A 4-port PNA-L. This VNA provides the combination of speed and accuracy for measuring multi-port and balanced components such as filters, duplexer, and RF modules up to 20 GHz. The N5230A VNA provides the following features and benefits:

- Full 4-port *S* parameter and balanced measurements up to 20 GHz
- 120 dB dynamic range at 20 GHz
- <0.006 dB of trace noise at 100 kHz intermediate frequency band width (IFBW)
- <4 μs/point measurement speed
- Automatic port extension automatically corrects for in-fixture measurements
- Advanced connectivity with LAN, universal serial bus (USB), and general purpose interface bus (GPIB) interfaces

4.4. Antennas

The same wide band antennas (Q-par Angus Ltd.) were used for transmission and reception in experiments performed between 2 and 18 GHz. The Q-par Angus Ltd. (model number WBH2-18HN/S) has a frequency range between 2 and 18 GHz, with a nominal gain between 10 and 22 dBi and a nominal beam width between 6° and 11°. The voltage standing wave ratio (VSWR) of the antenna is less than 2.5:1 (typically <2.0:1), and its cross polar is less than −17 dB. The antenna dimensions are 622 × 165 × 165 mm approximately, it weighs 2.7 kg and operates for temperatures between −40 and +70°C.

4.5. System operating at 60 GHz

In order to provide a comparative approach, an extension to the proposed system is added in order to sustain a frequency signal of 60 GHz. This is achieved using up-conversion and down-conversion methods. As the operational range of the VNA is limited to 20 GHz, a 3.5 GHz signal is up-converted to 60 GHz, then transmitted. The received 60 GHz signal is down-converted to 3.5 GHz. The measurement system is shown in [49]. The up- and down-conversion processes are obtained as follows: the VNA generates a CW signal at 3.5 GHz. Mixed with the phase locked oscillator (PLO) at 56.5 GHz frequency, the 3.5 GHz frequency is up-converted to 60 GHz. The IF signal is sent to the RF block. This block is composed of a mixer, a frequency tripler, a PLO at 18.83 GHz and a Band-Pass Filter (BPF) (59–61 GHz). The local oscillator (56.5 GHz) frequency is obtained with an 18.83 GHz PLO with 70 MHz external reference and a frequency tripler. The phase noise of the 18.83 GHz PLO signal is about −110 dBc/Hz at 10 kHz off carrier. The Band Pass Filter (BPF) with a bandwidth of 59–61 GHz removes out-of-band spurious signals caused by the modulator operation. The 0-dBm obtained signal is fed into a horn antenna with a gain of 22.4 dBi and a half-power beam width (HPBW) of 10° E and 12° H. The receiving antenna, identical to the transmitting horn antenna, is connected to a BPF (59–61 GHz). The input BPF removes the out-of-band noise. The RF filtered signal is down-converted to an IF signal centered at 3.5 GHz and fed into a BPF with a bandwidth of 2 GHz. A low noise amplifier (LNA) in the band of 2–4 GHz with a gain of 45 dB (noise factor 0.5 dB) is used to achieve sufficient gain. A variable attenuator with a dynamic range of 70 dB is used to control the IF power of IF input signal.

4.6. Preliminary results

The chest displacement varies between 4 and 12 mm due to respiration, while it ranges between 0.2 and 0.5 mm due to heart beating [19]. The measurement of this small displacement is the objective of this work. The variation of the phase of S_{21} is directly proportional to the chest displacement and indirectly proportional to the wavelength of the signal according to the following relationship:

$$\Delta\theta(t) = \frac{4\pi\Delta x(t)}{\lambda} \tag{3}$$

where λ is the wavelength of the transmitted signal and Δ is the chest displacement.

In order to validate the proposed system, measurements were performed at different frequencies. This section describes the measurements setup. As the frequency range of the antenna is between 2 and 18 GHz, and the maximum frequency of the VNA is 20 GHz, a set of experiments were performed between 2 and 18 GHz. Specifically, experiments were performed at 2.4, 5.8, 10, and 16 GHz. Another operational frequency, 60 GHz, was used via up- and down-conversion methods between 3.5 and 60 GHz. These two versions of the system were tested with a total output power of −10 dBm and for a distance of 1 m between the antennas and the person. The total output power, in other words radiated power, is the transmitted power added to the gain of the antenna. Measurements were performed on a 27-year-old healthy person, while holding the breath for a duration of 10 s [50]. The number of points taken for this window of time is 20,000 points. Hence, a sampling frequency of 2 kHz is obtained.

Performing a measurement begins by generating a continuous wave signal at the desired operational frequency. This CW signal, generated by the VNA, is driven to the transmitting antenna that is directed to the subject's chest. Reflected off the chest of the person under test, the signal is received by the receiving antenna and is driven back to the VNA. The phase variation S_{21}, which corresponds to the difference in terms of phase between the transmitted and the received signal, is computed. The difference in phase is due to the chest displacement. Hence, it contains information about the cardiopulmonary signals when breathing normally and about the heartbeat signal when holding the breath.

The theoretical values of the phase variation due to the chest displacement ranges and the average phase variation obtained by measurements when operating at 2.4, 5.8, 10, 16, and 60 GHz are shown in **Table 1**.

Frequency	Wavelength (λ)	$\Delta\theta$ for $\Delta x = 0.2$ mm	$\Delta\theta$ for $\Delta x = 0.5$ mm	Experimental $\Delta\theta$
2.4 GHz	125 mm	1.15°	2.88°	1.57°
5.8 GHz	51.72 mm	2.78°	6.96°	3.66°
10 GHz	30 mm	4.8°	12°	5.27°
16 GHz	18.75 mm	7.68°	19.2°	10.46°
60 GHz	5 mm	28.8°	72°	43.85°

Table 1. Theoretical and measured phase variations due to chest displacement.

For each of the utilized frequencies, the phase variation ranges within the theoretical limits. **Figure 2** shows the phase variations due to heartbeat signals detected at different frequencies and plotted within the same scale. It can be noticed that the phase variation increases when

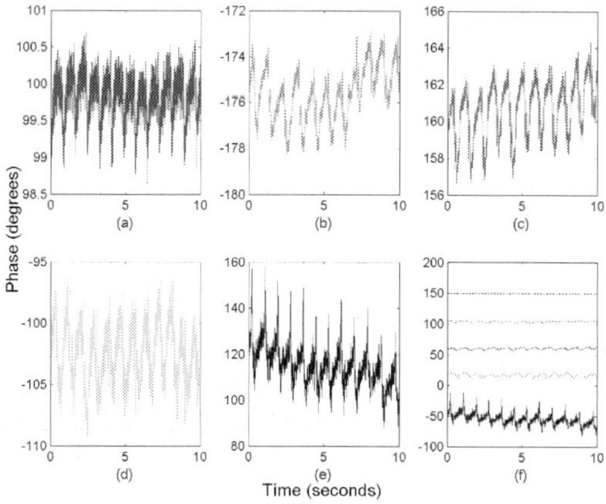

Figure 2. Phase variation of S_{21} due to heartbeat signal measured when holding the breath at different operational frequencies: (a) 2.4 GHz, (b) 5.8 GHz, (c) 10 GHz, (d) 16 GHz, (e) 60 GHz, and (f) all frequencies over the same scaling.

the frequency increases. Thus, higher sensitivity to small displacements is obtained at higher frequencies. The use of higher frequencies will reduce the noise effect and increase the accuracy in detecting the peaks of the signal.

5. Cardiopulmonary signals at different sides: front, back, left, and right

As the cardiopulmonary monitoring for patients and people under rubble requires heartbeat detection regardless their positions with respect to the system, measurements were performed at the four different sides from the PUT: front, back, left, and right. This section describes the measurement and shows the results of the contactless cardiopulmonary detection at four different sides.

5.1. Measurement setup

The measurements were performed on a 54-year-old healthy subject, sitting at a distance of 1 m from the antennas. The operational frequency of the microwave system is 5.8 GHz with a total output power of 0 dBm. Each measurement lasts 30 s where the PUT breathes normally. The contactless measurement signal is acquired simultaneously with a PC-based ECG to be used as a reference signal for the heartbeat detection in order to validate the accuracy of both the microwave system and the signal processing technique. **Figure 3** presents the phase of S_{21} measured at four positions: (a) measurement from the front side of the person, (b) measurement from the back side of the person, (c) measurement from the left side of the person, and (d) measurement from the right side of the person.

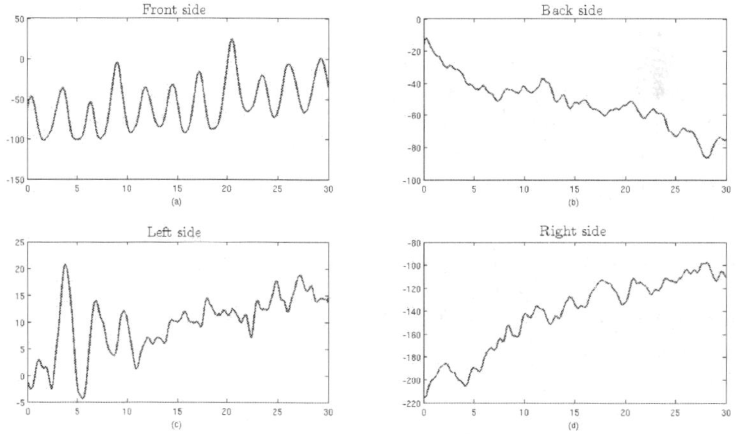

Figure 3. Phase variation of S_{21} due to the cardiopulmonary activities measurement at different sides from the subject: (a) front side, (b) back side, (c) left side, and (d) right side.

It can be noticed that the respiration signal is from the front side of the subject is clearer than other sides in time domain. The signal processing technique in the next subsection will show how it is possible to extract the heartbeat signal from the cardiopulmonary signal over all sides.

5.2. Signal processing techniques

Because the phase variations of S_{21} caused by respiration are larger than those caused by the heart beating, processing techniques are required to extract heartbeat signal from the obtained cardiopulmonary signal. Previous works tend to apply the FFT in order to extract the heartbeat rate. This gives an average value of the HR over a specific window of time; hence, it lacks providing information about the variation of the HR in time and cannot be established in real time as it needs a long-duration window. In order to overcome these problems, the discrete wavelet transform (DWT) is applied to extract the heartbeat signal.

The DWT (W_j, k) of a signal $f(t)$ is given by the scalar product of $f(t)$ with the scaling function (i.e. the wavelet basis function $\phi(t)$ which is scaled and shifted:

$$(W_j, k) \tag{4}$$

where the basis function is given by

$$\phi_j, k(t) = 2^{\frac{-j}{2}} \phi(2^{-j}t - k) \tag{5}$$

where j is the jth decomposition level or step and k is the kth wavelet coefficient at the jth level [51]. DWT is computed by successive low-pass and high-pass filtering of the discrete time-domain signal [51]. DWT determination examines the signal at different frequency bands with different resolutions by decomposing the signal into approximation coefficients (A) and detailed information (D). Hence, this algorithm gives precise analysis of frequency domain at low frequency and time domain at high frequency. The DWT principle is resumed in **Figure 4**.

In general, Dn contains frequencies between $fs/2^n$ and $fs/2^{n+1}$. As the HR varies between 60 and 120 beats per minute, the frequency of the heartbeat is located between 1 and 2 Hz. For the actual sampling frequency used in the VNA (666.7 Hz), no decomposition provides the signal having its frequency component between 1 and 2 Hz. Hence, a re-sampling is needed in order to convert the sampling frequency from 666.7 to 512 Hz. This lets the 1–2 Hz components be included in the 8th-level decomposition of the wavelet. Once the wavelet decomposition is extracted, the signal is reconstructed in time domain, and a peak detection method can be applied in order to detect peaks (beats), thus, to extract the heartbeat rate. The wavelet transform used in this study is the Bior2.4.

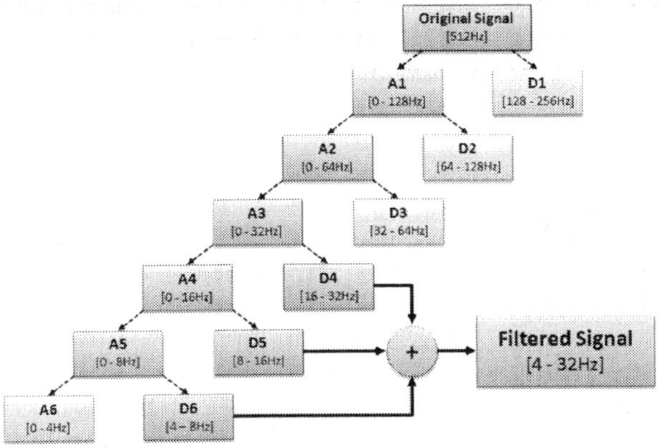

Figure 4. Example of wavelet decomposition.

Figure 5 shows the result of the 8th-level decomposition upon applying Bior2.4 to the cardiopulmonary signal detected from the front side of the PUT as well as the ECG signal extracted simultaneously with the cardiopulmonary signal.

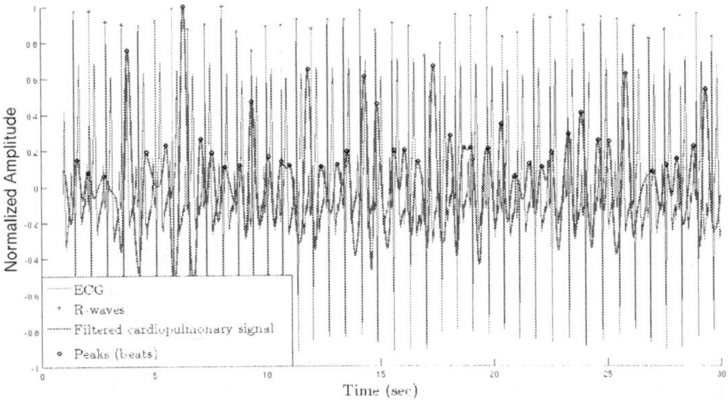

Figure 5. ECG vs. cardiopulmonary signal extracted from the front side of the subject and processed using DWT Bior2.4 level 8.

Applying the peak detection method to both ECG and filtered cardiopulmonary signal detected at the front side from the subject gives, respectively, 41 R-waves and 44 peaks. This results in an HR of 85 bpm for the ECG and 93 bpm for the filtered cardiopulmonary signal.

The result of the 8th-level decomposition upon applying Bior2.4 to the cardiopulmonary signal detected from the back side of the PUT is shown in **Figure 6**.

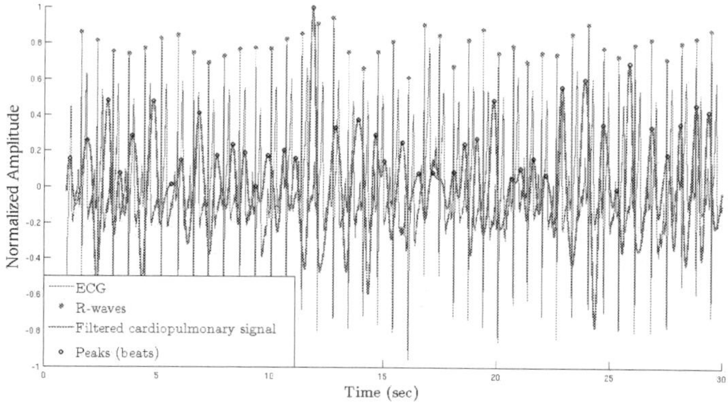

Figure 6. ECG vs. cardiopulmonary signal extracted from the back side of the subject and processed using DWT Bior2.4 level 8.

Applying the peak detection method to both ECG and filtered cardiopulmonary signal detected at the back side from the subject gives, respectively, 42 R-waves and 42 peaks for both signals. This results in an HR of 88 bpm for the ECG and 87 bpm for the filtered cardiopulmonary signal.

The result of the 8th-level decomposition upon applying Bior2.4 to the cardiopulmonary signal detected from the left side of the PUT is shown in **Figure 7**.

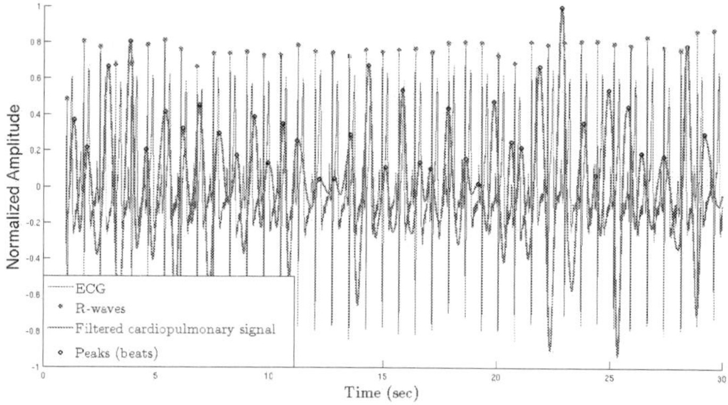

Figure 7. ECG vs. cardiopulmonary signal extracted from the left side of the subject and processed using DWT Bior2.4 level 8.

Applying the peak detection method to both ECG and filtered cardiopulmonary signal detected at the left side from the subject gives, respectively, 40 R-waves and 38 peaks. This results in an HR of 82 bpm for the ECG and 80 bpm for the filtered cardiopulmonary signal.

The result of the 8th-level decomposition upon applying Bior2.4 to the cardiopulmonary signal detected from the right side of the PUT is shown in **Figure 8**.

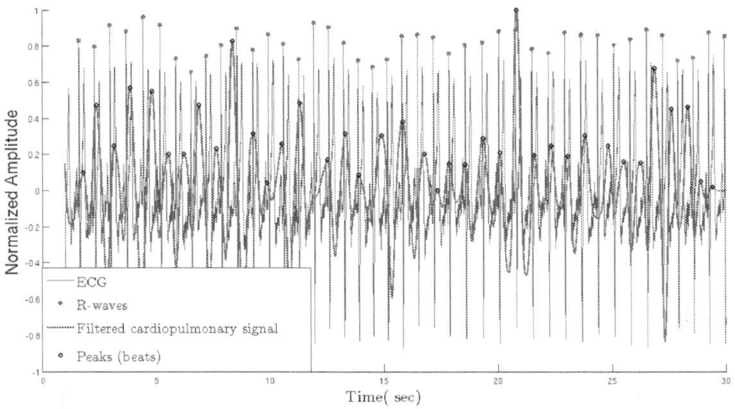

Figure 8. ECG vs. cardiopulmonary signal extracted from the right side of the subject and processed using DWT Bior2.4 level 8.

Applying the peak detection method to both ECG and filtered cardiopulmonary signal detected at the right side from the subject gives, respectively, 42 R-waves and 38 peaks. This results in an HR of 87 bpm for the ECG and 80 bpm for the filtered cardiopulmonary signal.

5.3. Results and discussion

Compared to the HR extracted from the ECG signal, the DWT with Bior2.4 family at decomposition level 8 shows accurate heartbeat detection. The signal processing technique applied to the signal extracted from the front side of the subject shows an error of 9%, while an error of 1% is obtained from the signal extracted at the back side of the subject. Also, the DWT applied to the signal extracted from the left side shows an error of 3% while an error of 7% is obtained for the signal extracted from the right side. **Table 2** shows the results in terms of HR calculation for both ECG and filtered cardiopulmonary signal for the four sides' measurements.

The heartbeat rate for both the ECG and VNA signals are calculated as follows:

$$HR = \frac{60(N-1)}{d_1 + d_2 + \ldots + d_{N-1}} \tag{6}$$

where N is the number of peaks and d_k is duration between two consecutive peaks. The peaks are detected using a classical peak-detection algorithm that is applied to both the ECG signal and the filtered cardiopulmonary signal.

Measurement side	Heartbeat rate for ECG signal (bpm)	Heartbeat rate for filtered cardiopulmonary signal (bpm)	Absolute relative error (%)
Front	85	93	9.12
Back	88	87	1.43
Left	82	80	2.66
Right	87	80	7.38

Table 2. HR calculation for ECG and filtered cardiopulmonary signals and the obtained relative error.

The relative error of the HR is calculated as

$$Error = \frac{100^* \mid HR_{ECG} - HR_{VNA} \mid}{HR_{ECG}} \tag{7}$$

As shown in Section 5.1, the cardiopulmonary signal detected from the front side of the subject shows clear respiration signal in time-domain while other sides' signals do not. On the other hand, when applying the Bior2.4 family of the discrete wavelet transform to the cardiopulmonary signals, the highest accuracy in terms of heartbeat rate is obtained at the back side while the front side shows the lowest accuracy. This is due to the fact that chest displacement due to breathing is higher at the front side than other sides; hence, the chest displacement due to heart beating will be less affected by the respiration signal on other sides.

6. Conclusion

A microwave system used in order to detect the chest wall motion that contains information about respiration and heart beating is described. The system is tested at different operational frequencies: 2.4, 5.8, 10, 16, and 60 GHz on a subject at 1 m from the system while holding the breath for 10 s. Other measurements were performed at 5.8 GHz for different positions for the subject: front, back, left, and right sides. The first measurement is performed on a 27-year-old subject while holding the breath, while the second measurement is performed on a 54-year-old subject while breathing normally. Along with a PC-based ECG, measurements are performed with 0 dBm output power and for a duration of 30 s where the subject breathes normally. The proposed system shows the ability of detecting cardiopulmonary signals for the four sides' positioning: front, back, left, and right. Wavelet transformation is used in processing cardiopulmonary signals in order to extract the heartbeat signal. The 8th-level decomposition of Bior2.4 shows high performance in providing the heartbeat signal in time domain where high accuracy is obtained in terms of heartbeat rate. A peak detection method is applied to the reconstructed signal from the 8th-level Bior2.4 decomposition. Accuracy in terms of HR for different positions varies between 1% (from the back side) and 9% (from the front side). The proposed system and signal processing technique show the possibility of measuring the cardiopulmonary activities of the subject at four different positions with high accuracy. Compared to other studies, the proposed processing technique shows the ability of detecting the heartbeat signal in time domain, hence, giving information about the variation of the heartbeat rate in real time with an error less than 10%. Future work will concern performing measurements on persons with different ages and under different breathing circumstances as well as for persons in motion.

Author details

Dany Obeid[1*], Sarah Samad[2], Sawsan Sadek[2], Gheorghe Zaharia[1] and Ghaïs El Zein[1]

*Address all correspondence to: dany.obeid@gmail.com

1 IETR UMR 6164, INSA Rennes, Rennes, France

2 University Institute of Technology, Lebanese University, Saida, Lebanon

References

[1] E. Braunwald and J. K. Perlkoff, Physical Examination of the Heart and Circulation, ser. Heart Disease: A Textbook of Cardiovascular Medicine, 2001.

[2] W. Dressler, Pulsations of the chest wall, 1937.

[3] P. M. S. Gillam, A. A. Deliyannis, and J. P. D. Mounsey, The left parasternal impulse, 1964.

[4] E. H. Awtry and J. Loscalzo, Evaluation of the Patient with Cardiovascular Disease, 2001.

[5] A. A. Deliyannis, P. M. S. Gillam, J. P. D. Mounsey, and R. E. Steiner, Thecardiac impulse and the motion of the heart, 1964.

[6] A. E. Aubert, L. Welkenhuysen, J. Montald, L. de Wolf, H. Geivers, J. Minten, H. Kesteloot, and H. Geest, Laser method for recording displacement of the heart and chest wall, 1984.

[7] C. M. Brandt, H. Annoni, J. Harthong, J. M. Reiner, and R. Krauskhopff, Evaluation of chest wall distortion related to cardiac activity by structured lights: A study of the apical impulse by the moiré technique, 1986.

[8] G. Ramachandran and M. Singh, Three-dimensional reconstruction of cardiac displacement patterns on the chest wall during the p, qrs, and t-segments of the ECG by laser speckle interferometry, 1989.

[9] G. Ramachandran, S. Swarnamani, and M. Singh, Reconstruction of out-of-plane cardiac displacement patterns as observed on the chest wall during various phases of ECG by capacitance transducer, 1991.

[10] K. Mohri, T. Jinnouchi, and K. Kawano, Accurate mechanocardiogram sensors using amorphous star-shaped core multivibrator combined with a magnet, 1987.

[11] K. Ikegaya, N. Suzumura, and T. Funada, Absolute calibration of phonocardiographic microphones and measurements of chest wall vibration, 1971.

[12] G. W. Yip, Y. Zhang, P. Y. Tan, M. Wang, P. Y. Ho, L. A. Brodin, and J. E. Sanderson, Left ventricular long-axis changes in early diastole: impact of systolic function on diastole, 2002.

[13] A. Owen, Effect of increasing age on diastolic motion of the left ventricular atrioventricular plane in normal subjects, 1999.

[14] O. X. Arcem, O. A. Knudson, M. C. Ellison, P. Baselga, D. D. Ivy, C. DeGroff, and L. Valdes-Cruz, Longitudinal motion of the atrioventricular annuli in children: reference values, growth-related changes, and effects of right ventricular volume and pressure overload, 2002.

[15] T. Kawasaki, S. Sasayama, S. I. Yagi, and T. A. T. Hirai, Non-invasive assessment of the age related changes in stiffness of major branches of the human arteries, 1987.

[16] J. M. Meinders and A. P. G. Hoeks, Simultaneous assessment of diameter and pressure waveforms in the carotid artery, 2004.

[17] T. Kondo, I. Kobayashi, Y. Taguchi, Y. Ohta, and N. Yanagimachi, Adynamic analysis of chest wall motions with MRI in healthy young subjects, 2000.

[18] T. A. Wilson, K. Rehder, S. Krayer, A. Hoffman, C. G. Whitney, and J. R. Rodarte, Geometry and respiratory displacement of human ribs, 1987.

[19] A. DeGroote, M. Wantier, G. Cheron, M. Estenne, and M. Pavia, Chest wall motion during tidal breathing, 1997.

[20] T. Kondo, T. Uhlig, P. Pemberton, and P. D. Sly, Laser monitoring of chest wall displacement, 1997.

[21] J. C. Lin, Non-invasive microwave measurement of respiration, 1975.

[22] K. H. Chan and J. C. Lin, Microprocessor-based cardiopulmonary rate monitor, 1987.

[23] H. R. Chuang, Y. F. Chen, and K. M. Chen, Microprocessor controlled automatic clutter-cancellation circuits for microwave systems to sense physiological movements remotely through the rubble, 1990.

[24] E. F. Greneker, Radar sensing of heartbeat and respiration at a distance with applications of the technology, 1997.

[25] K. M. Chen, Y. Huang, J. Zhang, and A. Norman, Microwave life-detection systems for searching human subjects under earthquake rubble or behind barrier, 2000.

[26] I. Arai, Survivor search radar system for persons trapped under earthquake rubble, 2001.

[27] A. D. Droitcour, O. B. Lubecke, V. M. Lubecke, and J. Lin, 0.25 μm CMOS and BiCMOS single chip direct conversion Doppler radars for remote sensing of vital signs, 2002.

[28] O. B. Lubecke, G. Awater, and V. M. Lubecke, Wireless LAN PC card sensing of vital signs, 2003.

[29] IEEE Standard for Safety Levels with Respect to Human Exposure to Radio Frequency Electromagnetic Fields, 3 kHz to 300 GHz Amendment 2: Specific Absorption Rate (SAR) Limits for the Pinna, 2004.

[30] Y. Xiao, J. Lin, O. B. Lubecke, and V. M. Lubecke, A ka-band low power Doppler radar system for remote detection of cardiopulmonary motion, 2005.

[31] Q. Zhou, J. Liu, A. H. Madsen, O. B. Lubecke, and V. Lubecke, Detection of multiple heartbeats using Doppler radar, 2006.

[32] O. B. Lubecke, V. Lubecke, A. Host-Madsen, D. Samardzija, and K. Cheung, Doppler radar sensing of multiple subjects in single and multiple antenna systems, 2005.

[33] V. M. Lubecke, O. B. Lubecke, A. H. Madsen, and A. E. Fathy, Through-the-wall radar life detection and monitoring, 2007.

[34] C. Li and J. Lin, Random body movement cancellation in doppler radar vital sign detection, 2008.

[35] Y. Wang; Q. Liu; A. E. Fathy, CW and pulse–Doppler radar processing based on FPGA for human sensing applications, 2012.

[36] W. Hu, Z. Zhao, Y. Wang, H. Zhang, and F. Lin, Noncontact accurate measurement of cardiopulmonary activity using a compact quadrature Doppler radar sensor, 2013.

[37] C. Li and J. Lin, Complex signal demodulation and random body movement cancellation techniques for non-contact vital sign detection, 2008.

[38] D. L. Gorgas, Vital Signs and Patient Monitoring Techniques, 2004.

[39] V. Lubecke, O. B. Lubecke, and E. Beck, A compact low-cost add-on module for Doppler radar sensing of vital signs using a wireless communications terminal, 2002.

[40] B. Lohman, O. B. Lubecke, V. Lubecke, P. Ong, and M. Sondhi, A digital signal processor for Doppler radar sensing of vital signs, 2002.

[41] C. Li, Y. Xiao, and J. Lin, Experiment and spectral analysis of a low-power ka-band heartbeat detector measuring from four sides of a human body, 2006.

[42] J. Li, L. Liu, Z. Zeng, and F. Liu, Advanced signal processing for vital sign extraction with applications in UWB radar detection of trapped victims in complex environments. IEEE Journal of Selected Topics in Applied Earth Observations and Remote Sensing. 2013;7(3):783–791.

[43] H. Zhang; S. Li ;X. Jing ;P. Zhang ;Y. Zhang, T. Jiao, G. Lu, J. Wang, The separation of the heartbeat and respiratory signal of a Doppler radar based on the LMS adaptive harmonic cancellation algorithm, 2013.

[44] M. K. Singh, A. Gupta, S. Permanand, U. Singh, O. P. Singh, Y. H. Kim, A. K. Singh, System modeling and signal processing of microwave Doppler radar for cardiopulmonary sensing, 2015.

[45] J. Tu, J. Lin, Fast acquisition of heart rate in noncontact vital sign radar measurement using time window-variation technique, 2015.

[46] IEEE Standard for Safety Levels with Respect to Human Exposure to Radio Frequency Electromagnetic Fields, 3 kHz to 300 GHz, 1992.

[47] J. C. Lin, Safety standards for human exposure to radio frequency radiation and their biological rationale, 2003.

[48] Biological effects and exposure criteria for radio frequency electromagnetic fields, 1986.

[49] D. Obeid, S. Sadek, G. Zaharia, and G. El-Zein, Non-contact heart-beat detection at 2.4, 5.8 and 60 GHz: a comparative study, 2009.

[50] D. Obeid, G. Zaharia, S. Sadek, G. El Zein, Microwave Doppler radar for heartbeat detection vs electrocardiogram, 2012.

[51] G. Alan Blackburn, J. Garke Ferwerda, Retrieval of chlorophyll concentration from leaf reflectance spectra using wavelet analysis, 2007.

[52] A. Ronaszeki, A. E. Aubert, and H. de Geest, Laser apexcardiogram in healthy young men: A comparative study with the conventional method, 1990.

[53] M. Singh and G. Ramachandran, Reconstruction of sequential cardiac inplane displacement patterns on the chest wall by laser speckle interferometry, 1991.

[54] K. Mohri, T. Kondo, H. Sugino, J. Yamasaki, and K. Yoshino, Non-contact linear displacement sensors using amorphous-core multivibrators for mechanocardiography, 1985.

[55] J. Lin, J. Kiernicki, M. Kiernicki, and P. Wollschlaeger, Microwave apexcardiography, 1979.

[56] Y. Xiao, J. Lin, Boric-Lubecke, and V. M. Lubecke, Frequency tuning technique for remote detection of heartbeat and respiration using low-power double-sided and transmission in ka-band, 2006.

[57] J. C. Lin, E. Dawe, and J. Majcherek, A noninvasive microwave apnea detector, 1977.

[58] H. R. Chuang, Y. F. Chen, and K. M. Chen, Automatic clutter-canceler for microwave life-detection systems, 1991.

[59] C. Li, J. Lin, and Y. Xiao, Robust overnight monitoring of human vital signs by a noncontact respiration and heartbeat detector, 2006.

[60] Guidelines for limiting exposure to time-varying electric, magnetic and electromagnetic fields (up to 300 GHz), 1998.

[61] J. M. Osepchuk and R. C. Petersen, Historical review of RF exposure standards and the international committee on electromagnetic safety (ICES), 2003.

[62] D. Obeid, S. Sadek, G. Zaharia, and G. El Zein, Multitunable microwave system for touchless heartbeat detection and heart rate variability extraction, 2010.

Chapter 3

Nonlinear Adaptive Signal Processing Improves the Diagnostic Quality of Transabdominal Fetal Electrocardiography

Radek Martinek, Radana Kahankova, Hana Skukova,
Jaromir Konecny, Petr Bilik, Jan Zidek and
Homer Nazeran

Additional information is available at the end of the chapter

http://dx.doi.org/10.5772/64068

Abstract

The abdominal fetal electrocardiogram (fECG) conveys valuable information that can aid clinicians with the diagnosis and monitoring of a potentially at risk fetus during pregnancy and in childbirth. This chapter primarily focuses on noninvasive (external and indirect) transabdominal fECG monitoring. Even though it is the preferred monitoring method, unlike its classical invasive (internal and direct) counterpart (transvaginal monitoring), it may be contaminated by a variety of undesirable signals that deteriorate its quality and reduce its value in reliable detection of hypoxic conditions in the fetus. A stronger maternal electrocardiogram (the mECG signal) along with technical and biological artifacts constitutes the main interfering signal components that diminish the diagnostic quality of the transabdominal fECG recordings. Currently, transabdominal fECG monitoring relies solely on the determination of the fetus' pulse or heart rate (FHR) by detecting RR intervals and does not take into account the morphology and duration of the fECG waves (P, QRS, T), intervals, and segments, which collectively convey very useful diagnostic information in adult cardiology. The main reason for the exclusion of these valuable pieces of information in the determination of the fetus' status from clinical practice is the fact that there are no sufficiently reliable and well-proven techniques for accurate extraction of fECG signals and robust derivation of these informative features. To address this shortcoming in fetal cardiology, we focus on adaptive signal processing methods and pay particular attention to nonlinear approaches that carry great promise in improving the quality of transabdominal fECG monitoring and consequently impacting fetal cardiology in clinical practice. Our investigation and experimental results by using clinical-quality synthetic data generated by our novel fECG signal generator suggest that adaptive neuro-fuzzy inference systems could produce a significant advancement in fetal monitoring during pregnancy and childbirth. The possibility of using a single device to leverage two advanced methods of fetal monitoring, namely noninvasive cardiotocography (CTG) and ST segment analysis (STAN) simultaneously, to detect fetal hypoxic conditions is very promising.

Keywords: adaptive signal processing, adaptive neuro-fuzzy inference system, fetal electrocardiogram, transabdominal monitoring, noninvasive cardiotocography, noninvasive ST segment analysis (STAN)

1. Introduction

The fetal electrocardiogram (fECG) is a recoding of the electrical activity of the fetal heart and provides clinically significant information about the physiological state of a fetus during pregnancy and labor. Early detection of hypoxic states (hypoxemia, hypoxia, and asphyxia) achieved by fECG signal monitoring can ensure the fetus' well-being during these stages. For greater detail please see [1–3].

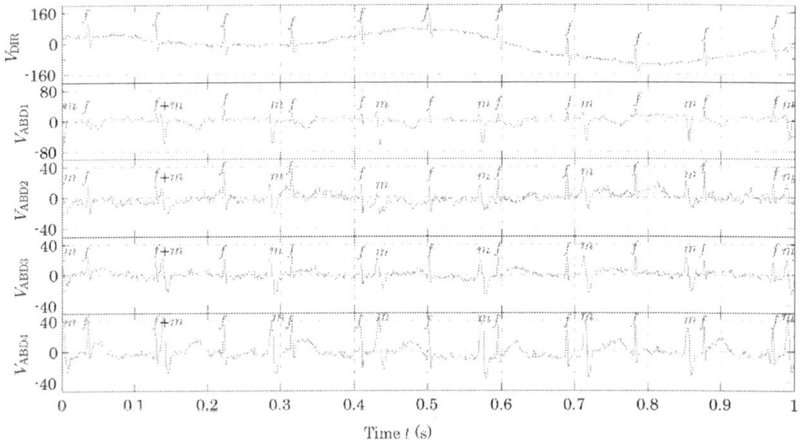

Figure 1. Real recordings of ECG signals by using the invasive and noninvasive techniques (f—fetal QRS, m—maternal QRS).

In clinical practice two methods are used to record fECG signals: invasive and noninvasive. The first one is direct and is performed transvaginally by using an Invasive Scalp Electrode (ISE). This approach is considered to be accurate as the fECG signals are recorded directly from the fetal's scalp without interference from the maternal heart (see **Figure 1**, the upper trace). However, it poses problems and risks to both the mother and the child (such as infections). In the noninvasive technique, multichannel skin bioelectrodes are placed on the mother's abdomen, and the simultaneous maternal (mECG) and fetal (fECG) signals, called the transabdominal or abdominal ECG (aECG), is acquired (See **Figure 1**, the lower 4 traces). This approach is convenient, noninvasive, and can be used during pregnancy and labor. However, there is a significant amount of overlap between fECG and mECG signals in addition to other undesirable signals such as bioelectric potentials (maternal muscle activity-mEMG, fetal movement activity, potentials generated by respiration and gastric activity, as well as power line interference [4,5], that deteriorate the quality of the afECG signals. **Figure 1** shows examples of fECG signals acquired by the invasive (V_{DIR}) and noninvasive (V_{ABD1}–V_{ABD4}) approaches.

We observe that the strong mECG and the weak fECG signals overlap in both time-domain (**Figure 1**) and frequency-domain (**Figure 2**). Therefore, filtering the mECG component from the composite aECG signal to produce diagnostic quality fECG is a very challenging signal processing task. Currently only a fraction of the vast amount of diagnostic information in the aECG is available to be used in clinical practice. Therefore, maximizing information extraction from aECG (in addition to cardiotocography—CTG) signals for the timely and reliable detection of fetal hypoxia is of tremendous clinical interest and could significantly impact the advancements in obstetrics.

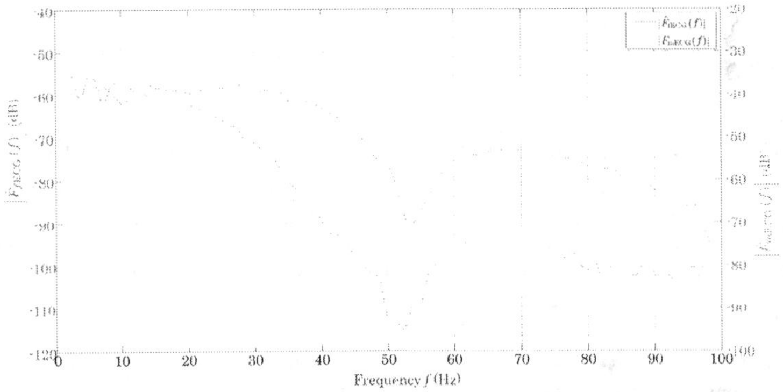

Figure 2. Abdominally recorded fECG and mECG in the frequency domain.

To address this signal processing challenge, many different approaches have been proposed to reliably detect fECG signals, but with varying degrees of success [4]. The holy grail of research in fetal electrocardiography is to fully recover the fECG signal and analyze its *morphology*, which produces valuable information on the fetus' status and health. The majority of recent techniques are mainly focused on the detection of the fetal heart rate (the intervals between R waves) with only a small portion being able to fully isolate the clinically useful ST interval and consequently perform accurate ST segment analysis (STAN) along with CTG. The only commercially available unit that has a built-in ability to perform STAN is Neoventa Medical's STAN S31. For a detailed description of this device, please see Ref. [6]. **Figure 3** shows an example of a real-time STAN. Fetal heart rate (fHR) and T/QRS are continuously displayed on the screen. These parameters are important in diagnosing hypoxic states. An increase in the ST segment and T wave as quantified by the ratio of the T wave to the QRS complex amplitude (T/QRS) has been associated with the different forms of the physiological responses expected in hypoxia (metabolic acidosis, myocardial glycogenolysis, etc.). For a more detailed explanation please see Reference [1].

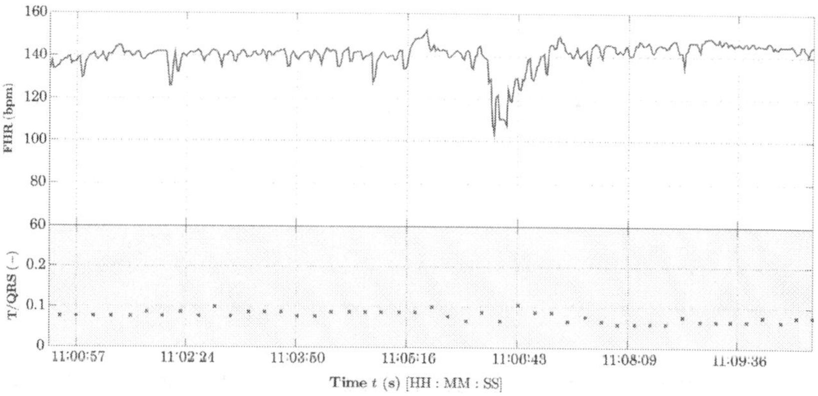

Figure 3. Real-time ST Analysis (Fetal Heart Rate, T wave and QRS complex ratio) using Matlab application.

A critical review of the current signal processing literature reveals that adaptive signal processing and soft computing methods are rapidly growing research areas and offer great promise to address some of the most challenging signal separation, pattern recognition, and classification problems in different areas of medicine including Obstetrics.

Driven by these advancements and promises, in the methods section we will first present a theoretical overview of the advanced signal processing (both nonadaptive and adaptive) techniques that have been applied to the extraction (separation) of fECG from aECG signals, and will choose a subset based upon their advantages. We will mainly focus on the Least Mean Squares (LMS) and Recursive Least Squares (RLS) algorithms [62]. Secondly, we will look at soft computing methods and describe how they have the ability to enhance the performance of adaptive signal processing algorithms in achieving better outcomes when processing biomedical signals. Then we will pay special attention to adaptive neuro-fuzzy inference systems (ANIFS) [60,63], which are considered to be the most significant in fetal electrocardiography research.

To provide a comparative analysis of the performance of our selected adaptive algorithms and their enhanced realizations using soft computing approaches, in the results section, we will report the outcomes of a number of experiments that we devised by using aECG (identical to clinical) signals generated by our novel LabVIEW-Based Multi-Channel Noninvasive Abdominal Maternal-Fetal Electrocardiogram Signal Generator [7,8]. This abdominal fECG signal generator allows us to realistically simulate all types of signal contaminations (both biological and nonbiological) affecting the quality of aECG signals.

Our experimental results were evaluated using both subjective and objective criteria. For the objective evaluation, we used the SNR values before and after processing, the RMSE value, and the required processing time for the selected data samples. These performance metrics (parameters) are defined in separate subsections of the methods section.

In conclusion, our experimental results using synthetically-generated (identical to clinical) data produced by our novel system revealed that it was possible to effectively extract fECG signals and significantly refine their diagnostic quality to enable reliable ST segment and CTG signal analysis. Refined aECG monitoring systems with built-in STAN and CTG analysis capabilities will pave the way for the timely and reliable detection of fetal hypoxia during pregnancy and labor, which is of tremendous clinical interest. These enhanced fECG monitoring systems will significantly impact the future advancements in Obstetrics.

2. Methods

2.1. fECG signal elicitation or extraction

Interference elimination can be implemented using a single-or a multi-channel source signal. These signals are then processed by various methods, which are used for fECG signal extraction, as shown in **Figure 4**. These methods are divided into two categories: nonadaptive and adaptive, depending on the system's inability or ability to accommodate unexpected changes.

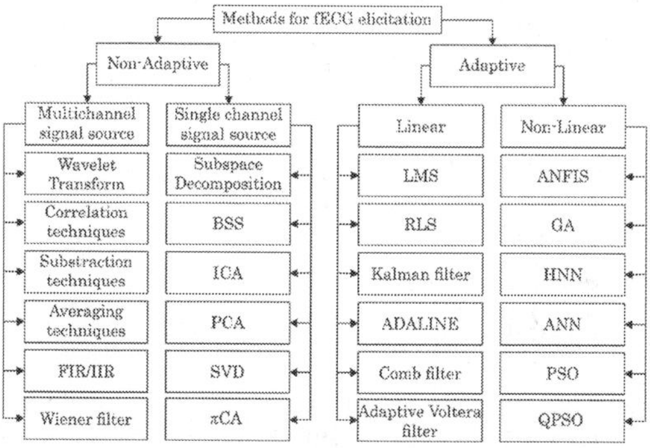

Figure 4. Summary of methods for fECG elicitation.

2.2. Nonadaptive methodologies

The Nonadaptive methodologies used for fECG signal extraction include Wavelet Transform-Based Techniques [9–11], Correlation Methods [12], Subtraction Methodologies [13], Single Value Decomposition (SVD) [14], Independent Component Analysis (ICA) and Blind Subspace Separation (BSS) [15–17], as well as Averaging Techniques [18].

The drawback of the nonadaptive techniques is that they are time-invariant in nature. Their time-invariance limitation has been overcome by the adaptive methods, which are more effective in reducing the overlapping noise (such as mECG) in time and frequency domains.

Nonadaptive methods are useful for data pre-processing or for noise elimination in case of classic ECGs [4].

2.3. Adaptive methodologies

Different variants of adaptive filters have been used for mECG signal cancellation and fECG signal extraction. These methods consist of training an adaptive or a matched filter for either removing the mECG signal using one or several maternal reference channels [19] or directly training the filter for extracting the fetal QRS complexes [20].

The existing adaptive filtering methods for maternal component elimination require a reference mECG channel that is morphologically similar to the contaminating waveform, or require several linearly independent channels to reconstruct any morphologic shape from the Ref. [21].

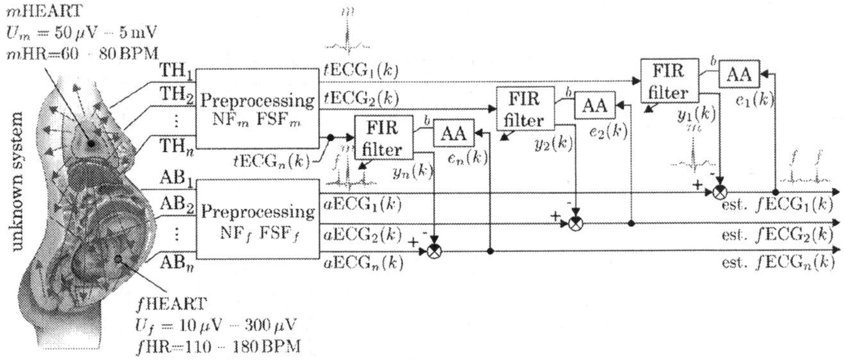

Figure 5. A theoretical multichannel adaptive noise (mECG and interferences) cancelation system.

Several approaches for mECG signal cancellation and fECG signal extraction have been used. The adaptive filters can be trained to extract the fetal QRS complexes directly or to estimate and remove the mECG component using reference maternal channels. The reference mECG signal can be recorded from the electrodes placed on the mother's thorax, or reconstructed from several abdominal channels that are linearly independent. The limitation of these approaches, which influences their performance, is that the morphology of the mECG signals highly depends on the electrode locations. Thus, the reconstruction of the complete ECG morphology from a linear combination of the reference electrodes is not always possible.

There are many different methodologies to extract fECG signals using adaptive filters based on one or several maternal reference channels (as shown in **Figure 5**). These methodologies include the LMS and RLS Algorithms, Artificial Intelligence (AI) Techniques, Fuzzy Inference Systems (FISs) [22,23], Genetic Algorithms (GA), and Bayesian Adaptive Filtering Frameworks which comprise Kalman Filters.

2.3.1. *Linear adaptive methods*

As mentioned before, adaptive methods can be linear or nonlinear. The linear methods for fECG signal extraction include algorithms such as LMS [24,25] RLS [24,26], Comb Filter [27], Adaptive Voltera Filter [28], Kalman Filter [29,30] or Adaptive Linear Networks (ADALINE) [31].

An adaptive filter is one that is characterized by the ability to self-adjust its coefficients according to an optimized training algorithm which is driven by a back-propagated error signal. Adaptive filters are used in noise cancellation applications to remove the noise adaptively from a signal and to improve the Signal to Noise Ratio (SNR) [4].

Simply said, it is a technique for the adaptive elimination of undesired signals (such as the maternal component) from the abdominal signal to obtain the fECG signal. The system can self-adjust to the existing circumstances and optimize its results.

2.3.2. *An example: an adaptive noise cancellation system for fECG signal extraction*

A theoretical multichannel adaptive noise cancellation system, shown in **Figure 5**, illustrates an adaptive elicitation technique of the fECG as an example. It consists of two kinds of input signals recorded from multiple leads: the abdominal ECG signals (AB_1–AB_n) and the thoracic ECG signals (TH_1–TH_n). Each abdominal signal consists of both maternal and fetal signals and serves as the primary input. The thoracic signal is considered to be completely maternal and is used as the reference input. Finite Impulse Response (FIR) Filter weights of the adaptive systems are updated by training algorithms based on the back-propagated error signal, which is the desired fECG signal ($fECG_1$–$fECG_n$). The maternal component is considered as noise to be eliminated. Each of the adaptive systems produces a signal, which is an approximation of the noise. This signal is subtracted from the abdominal ECG (aECG) signal so that the error signal that is back-propagated to the training algorithm is the fetal ECG signal with some noise.

Linear methods have limited performance in processing nonlinear or degenerate mixtures of signal and noise. In fact, fECG signals are not always linearly separable from undesirable signals contaminating them [32]. That is also a reason why linear algorithms yield better results when tested with synthetic data compared to those tested with real data. As the underlying physiological processes in the human body exhibit nonlinear behavior it seems more reasonable to use nonlinear methods for the construction of accurate and functional adaptive filters [22,32] to achieve better outcomes.

This chapter primarily focuses on the LMS- and RLS-based FIR Adaptive Filtering Methods. In the sections below we present mathematical descriptions for the most important methods such as LMS, Normalized LMS (NLMS), RLS, and Fast Transversal Filter (FTF).

2.4. Theoretical background

The Least Mean Squares (LMS) Algorithms are classified as adaptive filters that can change their coefficients to become a system that produces the least mean squares of the error (the difference between the desired and the actual) signal. It is a stochastic gradient descent method in that the filter is only adapted based on the error at the current time [33].

2.4.1. *Standard LMS*

The Standard LMS Algorithm performs the following operations to update the coefficients of an adaptive filter:

- Calculates the corresponding output signal from the adaptive filter by using the following equation:

$$y(k) = \sum_{i=0}^{N-1} w(k)x(k-i) = \mathbf{w}^{\mathrm{T}}(k)\mathbf{x}(k). \tag{1}$$

- Calculates the error signal $e(k)$ that denotes the difference between additional input signal $d(k)$ and $y(k)$ by using the following equation:

$$e(k) = d(k) - y(k). \tag{2}$$

- Updates the filter coefficients by using the following equation:

$$\mathbf{w}(k+1) = \mathbf{w}(k) + 2\mu e(k)\mathbf{x}(k), \tag{3}$$

where μ is the step size of the adaptive filter, is the filter coefficients vector, and $\mathbf{w}(k)$ is the input signal to a linear filter at time. Step size is a crucial parameter that can improve the convergence speed of the adaptive filter. It determines both how quickly and how closely the adaptive filter converges to the filter solution [34].

2.4.2. *Normalized LMS (NLMS) algorithm*

The NLMS Algorithm is a modified form of the standard LMS Algorithm. The NLMS Algorithm updates the coefficients of an adaptive filter by using the following equation:

$$\mathbf{w}(k+1) = \mathbf{w}(k) + \mu e(k)\frac{\mathbf{x}(k)}{\mathbf{x}(k)\;^{2}}. \tag{4}$$

It is obvious that the NLMS Algorithm is almost identical to the Standard LMS Algorithm except that the NLMS Algorithm has a time-varying step size $\mu(k)$, [34].

2.4.3. *The recursive least square (RLS) algorithm*

Unlike the LMS Algorithm, which reduces the mean square error, the principle of the RLS Algorithm is that it recursively finds the coefficients that minimize a weighted linear least squares cost function relating to the input signals. In the case of the RLS Algorithm, the input signals are considered deterministic, while for the LMS Algorithm, they are considered stochastic. Compared to most of its competitors, the RLS Algorithm exhibits extremely fast convergence. However, this benefit comes at the cost of high computational complexity.

The standard RLS Algorithm performs the following operations to update the coefficients of an adaptive filter:

- Calculates the output signal of the adaptive filter:

$$y(k) = \mathbf{w}^{\mathrm{T}}(k-1)\mathbf{x}(k). \tag{5}$$

- Calculates estimation error $e(k)$ by using the following equation:

$$e(k) = d(k) - y(k). \tag{6}$$

- Updates the filter coefficients by using the following equation:

$$w(k+1) = \mathbf{w}^{\mathrm{T}}(k) + e(k)\mathbf{K}(k), \tag{7}$$

where $\mathbf{w}(k)$ is the filter coefficients vector and $\mathbf{K}(k)$ is the gain vector and is defined by the following equation:

$$\mathbf{K}(k) = \frac{P(k)\mathbf{u}(k)}{\lambda + \mathbf{u}^{\mathrm{T}}(k)P(k)\mathbf{u}(k)}. \tag{8}$$

$P(k)$ is the inverse correlation matrix of the input signal. $P(k)$ has the following initial value:

$$P(k) = \begin{bmatrix} \delta^{-1} & 0 & \dots & 0 \\ 0 & \delta^{-1} & \dots & 0 \\ \dots & \dots & \dots & \dots \\ 0 & 0 & \dots & \delta^{-1} \end{bmatrix}, \tag{9}$$

where δ is the regularization factor. The standard RLS Algorithm uses the following equation to update this inverse correlation matrix.

$$P(k+1) = \lambda^{-1}P(k) - \lambda^{n-1}\mathbf{K}(k)\mathbf{u}^{\mathrm{T}}(k)P(k). \tag{10}$$

- Repeats all the steps for the next iteration $(k + 1)$.

The selection of the forgetting factor λ depends on the number of the samples k as follows:

$$k = \frac{\lambda}{1-\lambda}. \tag{11}$$

If the analyzed signal is stationary, λ should be chosen as unity. Otherwise, λ should be smaller than the unity to track the nonstationary portion of the signals. The performance index takes into account the most recent errors as calculated by the most recent k^{th} iteration [35].

2.4.4. Fast transversal filter (FTF)

The complexity of the classical RLS Algorithm (related to the speed of convergence) was the main reason for the inclusion of the FTF (Fast Transversal Filter) Algorithm. A detailed description of the algorithm is very extensive, complicated, and beyond the scope of this chapter. Its detailed formula derivation can be found in Ref. [36].

2.5. Soft computing methods

Biological systems including the human body and most of the real-world physical systems are highly involved nonlinear dynamical systems with a large degree of variability, imprecision, and uncertainty. As such, it is impossible for humans to find tractable solutions to problems associated with these systems without using powerful computing technologies. This is mainly attributable to the extensive amount of required data and processing time. In general, a nonlinear system is capable of generating quantitative or qualitative information. Quantitative information is represented by accurate numerical values, which are acquired by conventional modeling and mathematics. These conventional methods of data acquisition are rigorous and their results have to be precise, certain or categorically true or false. Precision and certainty of calculations are attainable at a higher computational cost. On the contrary, qualitative information contains knowledge or experience that can be expressed in natural language (e.g., big, medium, small). Qualitative information is processed by "soft" approaches called soft computing or artificial intelligence. Soft computing is a collection of methodologies that are able to tolerate imprecision and uncertainty and exploit these attributes to achieve robust and low-cost solutions. These methods aim to imitate the way the human brain processes information.

It is very difficult to describe real-world systems by classical mathematics, and it has been established that processing purely quantitative information is not efficient and represents a huge computational burden. New research has shown that nonlinear systems are substantially better modeled by artificial intelligence. These facts have led to the development of new intelligent soft computing methods. In current practice, these methods find various applications in software engineering, signal processing, and optimization. Soft computing methods comprised of fuzzy logic, artificial neural networks, evolutionary algorithms, and hybrid algorithms, are distinguishable from other computational techniques by exhibiting a tolerance for imprecision and uncertainty. Another advantage of soft computing methods is in their ability to adapt and learn, which makes them very suitable for adaptive filtering applications

http://dx.doi.org/10.5772/64068

where their training algorithms allow them to adapt the system's parameters to existing conditions.

The application of soft computing methods in the processing of fECG signals is still in its infancy. However, the rapid development of computing processor technology and computational intelligent algorithms in the last three decades has provided new impetus for advancing fetal cardiology. To date, a number of soft computing-based approaches have been introduced to tackle fECG signal extraction. These include Adaptive Linear Neural Network (ADALINE), [8,37], Genetic Algorithms (GA), [20,21], ANFIS [32,34,38], and ANFIS trained with Particle Swarm Optimization (PSO) [1,2]. In [37] authors used an ADALINE trained with the LMS Algorithm for suppression of mECG signals. The ADALINE was trained to eliminate the maternal component from the aECG. This was carried out by subtraction of the mECG from the aECG. The resultant error signal was equal to the fECG. Neural networks were also used in [8], in combination with FIR filters. Taking advantage of the combined capabilities of soft computing methods and digital FIR or IIR (Infinite Impulse Response) filters is now common. For example, in [20] authors proposed to apply low pass FIR filtering optimized by a Genetic Algorithm (GA). The GA-modified coefficients of the FIR filter produced the best possible results for fECG signal extraction. A comparison of the results yielded by this method with those produced by methods using the LMS and NLMS Algorithms showed that the quality of filtering using the GA with eight bits and ten iterations was equal to those of the other methods. Better results were achieved by using IIR instead of FIR filters [21]. An ANFIS tuned by PSO was considered to be an efficient tool for the extraction of not only the QRS complexes, but also all the components of the fECG signal. This level of performance has not been achieved by leveraging any other two leading methods to date. With this overview in mind, in the section below we describe the extraction of fECG signals by using ANFIS.

2.5.1. ANFIS theoretical background

An Adaptive Neuro-fuzzy System (ANFIS) is a hybrid adaptive network based on a Sugeno-type fuzzy interference system (FIS) implemented into a feed-forward artificial neural network framework [39–44]. It uses a neuro-adaptive learning algorithm to determine the relationship between the input and output data sets. This learning algorithm can be hybrid or use back propagation. The advantage of an ANFIS lies in its ability to combine the "cleverness" of the Artificial Neural Networks (ANNs) and Fuzzy Inference Systems (FISs) in learning nonlinearities, which complement each other. A FIS incorporates human knowledge into the system in contrast to an ANN, which is capable of optimizing the ANFIS' parameters in implementing the learning process. To ensure correct and smooth running of the ANFIS, a number of fundamental considerations has to be made:

- The ANFIS should be single output.
- The FIS has to be the Sugeno model of zero or first order.
- The number of rules should correspond to the number of membership functions.
- The output membership function should be constant or linear.

2.5.2. *ANFIS architecture*

The original Jang's ANFIS architecture consists of five feed-forward interconnected layers, namely: a fuzzy layer, a product layer, a normalized layer, a de-fuzzification layer, and a total output layer [45]. In each layer several nodes are included and described by the node function. The nodes in these layers have an adaptive or a fixed nature and the difference between them is shown graphically in **Figure 6**, in which circles indicate the fixed nodes whereas squares represent the adaptive ones. The elementary ANFIS architecture has two initial inputs and one total single-value output. The rule base of the Sugeno FIS model is constituted by two IF-THEN rules in the following form [45]:

$$\text{IF}(x \,\text{is}\, A_1)\text{and}(y \,is\, B_1)\text{THEN}(f_1 = p_1x + q_1y + r_1). \tag{12}$$

$$\text{IF}(x \,\text{is}\, A_2)\text{and}(y \,is\, B_2)\text{THEN}(f_2 = p_2x + q_2y + r_2). \tag{13}$$

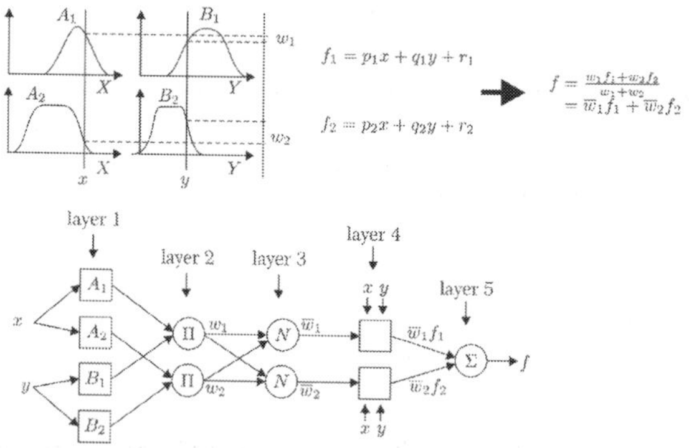

Figure 6. Fundamental scheme of ANFIS architecture.

where x and y are initial inputs; A_i and B_i are the nonlinear fuzzy sets also called a remise section; f_i is the output of the system; and p_i, q_i, and r_i are linear design parameters, which are determined during the training process.

Layer 1: The first layer of this architecture is an adaptive layer used for fuzzification of input variables. Each node represents the input value of a linguistic variable. The node function associated with the output of each node is

$$O_{1,i} = \mu_{A_i}(x); \text{for}\, i = 1,2., \tag{14}$$

$$O_{1,i} = \mu_{B_i-2}(y); \text{for}\, i = 3,4, \tag{15}$$

where x (or y) are inputs of the node i, A_i (or B_{i-1}) are linguistic labels and $\mu_{Ai}(x)$, respectively; μ_{Bi-2} (y) can accept any fuzzy membership function. In conclusion, $0_{1,i}$ is an expression of the membership function, in other words a membership grade, which indicates how much given x (or y) satisfies quantifier A_i (or B_i). The membership function can acquire several shapes including bell-shaped, triangular, and trapezoidal or Gaussian. For illustration we will use a bell-shaped Membership Function (MF) (Eq. 16).

$$\mu_{A_i}(x) = \frac{1}{1+\left|\frac{(x-c_i)}{a_i}\right|^{2b_i}} . \mu_{A_i}(x) = \frac{1}{1+\left|\frac{(x-c_i)}{a_i}\right|^{2b_i}} . \tag{16}$$

Parameters a_i, b_i, and c_i in Eq. (16) change the shape of the MF degree. Its value ranges from 0 to 1, where 0 is equal to the minimum value and 1 is equal to the maximum value.

Layer 2: The nodes in the second layer multiply the output signals from the previous layer. The output of this layer denotes $O_{2,i}$ and is described as:

$$O_{2,i} = w_i = \mu_{Ai}(x)\mu_{Bi}(y); \text{ for } i = 1,2. \tag{17}$$

Layer 3: The normalized layer labeled N contains a function to calculate the normalized firing strength. The output is labeled $O_{3,i}$.

$$O_{3,i} = \overline{w}_i = \frac{w_i}{w_1 + w_2}; \text{ for } i = 1,2. \tag{18}$$

Layer 4: All nodes in this layer are adaptive. The node function has the form given below:

$$O_{4,i} = \overline{w}_i f_i = \overline{w}_i \left(p_i x + q_i y + r_i\right), \text{for } i = 1,2, \tag{19}$$

where output $O_{4,i}$ defines a de-fuzzified (crisp) relationship between the input and output of this layer, is a firing strength desired in the normalized layer p_i, q_i, and r_i are linear adaptive parameters also called consequent parameters.

Layer 5: The last and fixed layer calculates the total output of the system.

$$O_{5,i} = \sum_i \overline{w}_i f_i = \frac{\sum_i \overline{w}_i f_i}{\sum_i \overline{w}_i}. \tag{20}$$

The output $O_{5,i}$ is derived from the summation of incoming signals from Layer 4.

2.5.3. Hybrid learning algorithm

The hybrid learning algorithm in the ANFIS tunes the parameters of the Sugeno type FIS. It is a combination of the LMS and the Back-propagation Gradient Descent Algorithm (BPG). Each part of the hybrid algorithm is focused on a different part of the rule base in the ANFIS architecture. The premise (antecedent) parameters are adjusted by the BPG Algorithm and the consequent parameters are tuned by the LMS Algorithm. With respect to this distribution, the hybrid learning algorithm is divided into two passes, which are regularly repeated with every epoch. These passes are called "forward" and "backward" passes. **Figure 7** depicts the scheme of the hybrid learning algorithm. This hybrid algorithm converges much faster than the original pure back-propagation algorithm, as the latter reduces the search space dimensions [46,47].

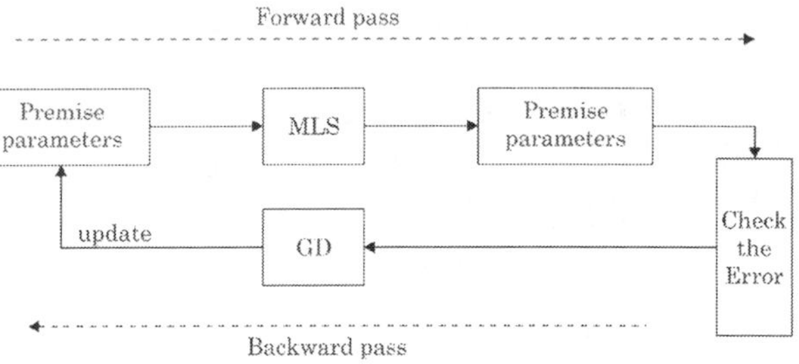

Figure 7. Block diagram for the hybrid-learning algorithm.

2.6. Definition of the parameters

The measurement of the quality of the fECG extraction procedures is based on the absence of noise and the degree of similarity between the recovered fECG signals and the ideal fECG signals, where the main parameters can be helpful to control the effectiveness of the fECG extraction and Signal to noise ratio (SNR).

2.6.1. Signal to noise ratio (SNR)

The relation between signal and noise is described by the SNR. To evaluate the filtering quality by the SNR, it is essential to calculate this ratio before and after filtering. The SNR before filtering is labeled SNR_{IN} and the SNR after filtrating is labeled SNR_{OUT}. Based on SNR_{IN} and SNR_{OUT}, it is possible to track the improvement of the SNR after filtering. Their expressions are as follows:

$$\mathrm{SNR_{IN}} = 10 \cdot \log_{10} \log_{10} \left(\frac{\sum_{i=1}^{N-1} \left[\mathrm{sig_{org}}(i) \right]^2}{\sum_{i=1}^{N-1} \left[\mathrm{sig_{noise}}(i) - \mathrm{sig_{org}}(i) \right]^2} \right), \tag{21}$$

where sig_{org} is the desired signal equal to an ideal fECG and sig_{noise} is a disturbing noise. This signal corresponds to a simulated mECG after passing through the unknown environment of the human body. Please clarify this sentence. Since the disturbance noise is a sum of the ideal fECG and mECG after they pass through the human body, it is necessary to subtract these two signals from each other in the denominator, SNR_{OUT} defines:

$$\mathrm{SNR}_{\mathrm{OUT}} = 10 \cdot \log_{10}\left(\frac{\sum_{i=1}^{N-1}\left[\mathrm{sig}_{\mathrm{org}}(i)\right]^2}{\sum_{i=1}^{N-1}\left[\mathrm{sig}_{\mathrm{rec}}(i) - \mathrm{sig}_{\mathrm{org}}(i)\right]^2} \right), \tag{22}$$

where sig_{org} denotes the original signal (ideal fECG) and sig_{rec} (i) the signal recovered by the algorithm.

It is possible to evaluate the effectiveness of the proposed adaptive method by finding the difference between SNR_{IN} and SNR_{OUT} .

The SNR quantifies the relation between the fetal ECG signal and the rest of the undesired components (mECG). In the general fECG inverse problem, this is not an operative definition of the SNR, since it requires knowing the contribution of the fetal ECG signal and the noise. Since our signals are synthetic, this information is available.

2.6.2. Mean square error (MSE) and root mean square error (RMSE)

The primary statistical measure used is a mean or a squared prediction error function. This evolved into widespread use of the mean squared prediction error as a performance measure, often shortened to simply the mean square error (MSE). It is a useful tool used for an evaluation of prediction, which reflects the degree of inaccuracy between an estimated and an original output described by:

$$MSE = \frac{1}{n}\sum_{i=1}^{n}\left(\mathrm{sig}_{\mathrm{rec}}(i) - \mathrm{sig}_{\mathrm{org}}(i)\right)^2, \tag{23}$$

where sig_{org} denotes the original signal (ideal fECG) and sig_{rec} the signal recovered by the algorithm.

MSE is often replaced by RMSE defined by:

$$RMSE = \frac{1}{n}\sqrt{\sum_{i=1}^{n}\left(\mathrm{sig}_{\mathrm{rec}}(i) - \mathrm{sig}_{\mathrm{org}}(i)\right)^2}. \tag{24}$$

where sig_{org} denotes the original signal (ideal fECG) and sig_{rec} the signal recovered by the algorithm.

RMSE is a measure of the differences between values predicted by a model or an estimator and the values observed. The closer this value is to zero the more accurate is the system.

2.7. Generation of test data form experiments

To objectively assess the results of fECG signal extraction approaches by using the quality metrics defined above (SNR, RMSE, and others), knowledge of the reference signals (mECG and ideal fECG) is essential. That is not possible in case of the clinical (real) data (as the reference fECG signal is missing). On the other hand, most commonly used synthetic data are often too idealized and do not include the influence of the nonlinear environment of the human body. That is the reason why most fECG signal extraction methods, which are considered successful when tested with idealized synthetic data do not produce useful results when tested with clinical (real) data.

Acknowledging the limitations associated with idealized synthetic data used in testing fECG signal processing algorithms, we took advantage of our novel fECG signal generator [7,8] to generate clinically realistic data [42] for our experiments. Our fECG signal generator is unique in many respects. It is designed to simulate the fetal heart activity while special attention is given to the fetal heart development in relation to the fetus' anatomy, physiology, and pathology. The noninvasive signal generator enables many parameters to be set, including Fetal Heart Rate (fHR), Maternal Heart Rate (mHR), Gestational Age, fECG interferences (biological and technical artifacts), as well as other fECG signal characteristics. Furthermore, based on the change in the fHR and in the T wave-to-QRS complex ratio (T/QRS), the generator enables manifestations of hypoxic states (hypoxemia, hypoxia, and asphyxia) to be monitored while complying with clinical recommendations for classifications in cardiotocography (CTG) and fECG ST segment analysis (STAN).

As described in detail elsewhere [7,8], the generator can produce realistic synthetic signals (identical to those acquired in clinical practice) with pre-defined properties for as many input as desired (n), see **Figure 5**. Such signals are well suited to the testing of existing and new methods of fECG processing [22,23,42].

The experiments were realized with six inputs, i.e., four channel combinations ($TE_2 \leftrightarrow BA_1$; $TE_2 \leftrightarrow BA_2$; $TE_2 \leftrightarrow BA_3$; $TE_1 \leftrightarrow BA_4$), which were processed collaterally by four independent adaptive systems.

For all the analyzed methods, the same starting parameters were selected according to clinical recommendations for CTG and STAN evaluations [42] as follows:

- record duration t = 03:00 (mm:ss), sample rate fs = 1 kHz, quantization step 0.1 mV,
- ideal mECG from thoracic electrodes TH_1 and TH_2 with variable MHR in the 65–85 bpm range,
- ideal fECG from abdominal electrodes AB_1, AB_2, AB_3 and AB_4 with the FHR in the 110–150 bpm range and the T/QRS complex in the 0.05–0.15 mV range,
- unwanted interference created in the human body modeled by empirical nonlinear transformation,

- SNR in for individual lead combinations and TH_1–AB_1 = −16.00 dB; TH_2–AB_3 = −27.70 dB; TH_1–AB_2 = −33.01 dB; TH_2–AB_4 = −23.35 dB,
- gestational age of fetus = 40 weeks (this parameter influences the length of each element of the fECG),
- for all of the adaptive algorithms used in the experiments the filter length was $N = 32$.

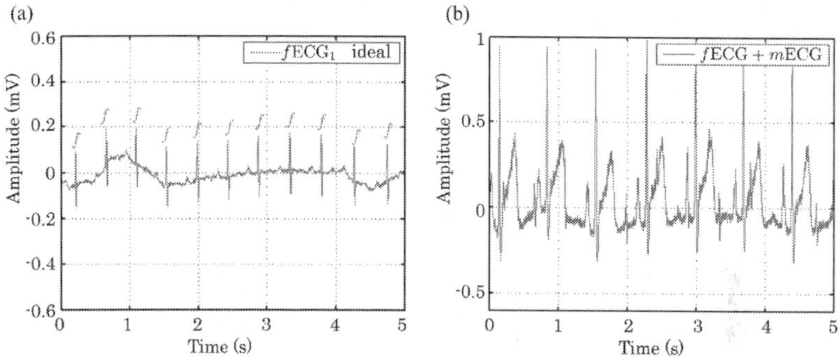

Figure 8. (a) Ideal fECG signal modelled by generator, (b) noisy fECG recordings modelled by generator.

To implement the required computational complexity, all experiments were performed by using a Personal Computer (PC) with a 3 GHz quad-core processor and 4 GB of RAM, please see Ref. [6]. **Figure 8a** shows the waveforms of the ideal and (**Figure 8b**) noisy fECG signals for the channel combination $TE_1 \leftrightarrow BA_1$, which were generated using the software-controlled generator.

3. Results and discussions

3.1. Experimental results: testing linear adaptive algorithms

The test signals generated by our generator were first processed by the LMS, NLMS, RLS and FTF Algorithms. The ideal combination of the selected settings for these functions was based on the input and output Signal to Noise Ratios (SNRs) as well as Root Mean Square Error (RMSE).

Figures 9 and **10** show the outputs of the adaptive systems for each one of the algorithms tested for the signals recorded by channels $TE_1 \leftrightarrow AE_1$. **Figure 9a** shows results of filtering aECG signals using the LMS algorithm and **Figure 9b** using NLMS algorithm. The LMS algorithm provides better results than NLMS algorithm; however, NLMS algorithm is more computational (processing time) expensive (see **Table 1**).

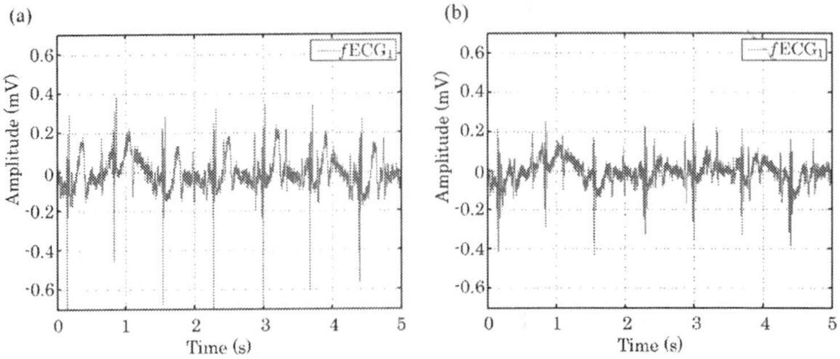

Figure 9. Output waveforms—results of filtering aECG signals (a) using the LMS and (b) using NLMS algorithms.

Name of adaptive filter used		Electrode combination	SNR_{in} (dB)	SNR_{out} (dB)	SNR_{imp} (dB)	RMSE (-)	Processing time (min:s)
LMS-based filters	LMS	$TE_1 \leftrightarrow AE_1$	-16.0036	-11.1935	4.8101	0.1388	00:57
		$TE_2 \leftrightarrow AE_2$	-33.0132	-32.1204	0.8928	0.4273	
		$TE_2 \leftrightarrow AE_3$	-27.7029	-21.1847	6.5182	0.1997	
		$TE_1 \leftrightarrow AE_4$	-23.3521	-22.2700	1.0821	0.3125	
	NLMS	$TE_1 \leftrightarrow AE_1$	-16.0036	-9.3647	6.6389	0.0996	01:15
		$TE_2 \leftrightarrow AE_2$	-33.0132	-28.0695	4.9437	0.2283	
		$TE_2 \leftrightarrow AE_3$	-27.7029	-20.3672	7.3357	0.1823	
		$TE_1 \leftrightarrow AE_4$	-23.3521	-19.5769	3.7792	0.2340	
RLS-based filters	RLS	$TE_1 \leftrightarrow AE_1$	-16.0036	-5.5187	10.4849	0.0843	01:49
		$TE_2 \leftrightarrow AE_2$	-33.0132	-25.5992	7.4140	0.1379	
		$TE_2 \leftrightarrow AE_3$	-27.7029	-17.6624	10.0405	0.1196	
		$TE_1 \leftrightarrow AE_4$	-23.3521	-15.8197	7.5324	0.1078	
	FTF	$TE_1 \leftrightarrow AE_1$	-16.0036	-6.2045	9.7991	0.0889	01:28
		$TE_2 \leftrightarrow AE_2$	-33.0132	-25.5174	7.4985	0.2146	
		$TE_2 \leftrightarrow AE_3$	-27.7029	-18.2309	9.4720	0.1794	
		$TE_1 \leftrightarrow AE_4$	-23.3521	-16.2830	7.0691	0.1617	

Table 1. Experimental results for LMS, NLMS, RLS, and FTF adaptive algorithms using synthetic fECG and mECG signals generated by our novel simulator.

Figure 10a shows results of filtering aECG signals using the RLS algorithm and **Figure 10b** using FTF algorithm. The RLS algorithm provides a bit better than FTF algorithm, nevertheless RLS algorithm is more computational expensive than RLS algorithm (see **Table 1**).

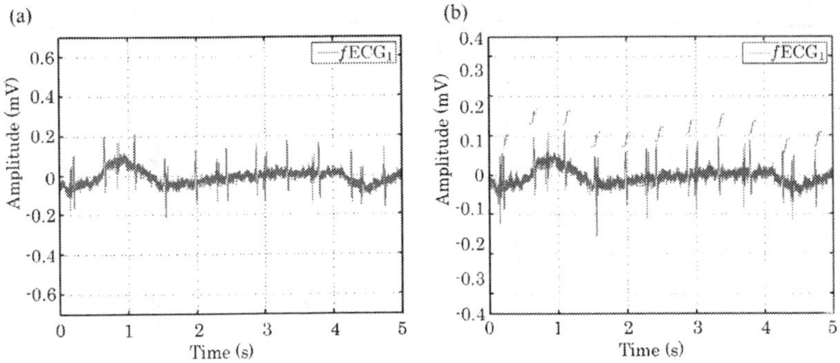

Figure 10. Output waveforms—results of filtering aECG signals (a) using the RLS and (b) using FTF algorithms.

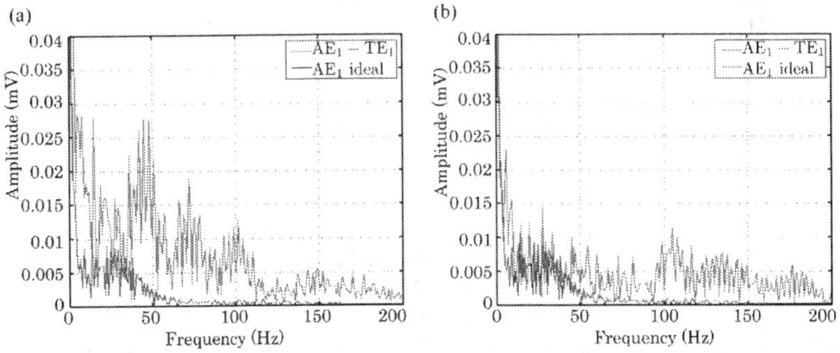

Figure 11. Amplitude spectrum—results of filtering aECG signals (a) using the LMS and (b) using NLMS Algorithms.

Figures 11 and **12** show the amplitude spectrums of the ideal fECG and the spectrum of the adaptive system output. **Figure 11a** shows amplitude spectrum of filtering aECG signals using the LMS algorithm and **Figure 11b** using the NLMS algorithm.

Figure 12a shows amplitude spectrum of filtering aECG signals using the RLS algorithm and **Figure 12b** using the FTF algorithm.

Table 1 shows the experimental results for the adaptive algorithms. The results presented in **Table 1** and **Figures 9–12** show some improvements in the frequency and time domains as measured by the quality parameters SNR and RMSE. The experimental results revealed that the RLS-based filters (RLS—**Figures 10a** and **12a** and FTF—**Figures 10b** and **12b**, Algorithms) produced the best outcomes. The computational times (1:49 and 1:28 s, respectively) for these algorithms were longer due to their complexity.

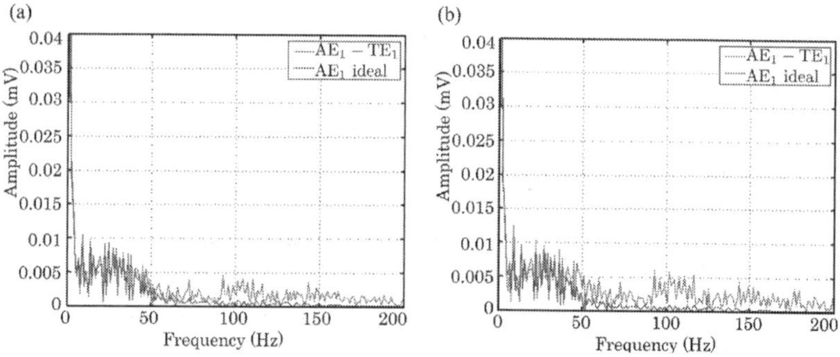

Figure 12. Amplitude spectrum—results of filtering aECG signals (a) using the RLS and (b) using FTF Algorithms.

Our experimental results described above indicate that the morphological differences between the original (ideal) and the recovered fECG signals are so significant that such processed signals cannot be satisfactorily used to detect fetal hypoxic conditions. These differences are mainly attributable to nonlinearities in the human body model. Therefore, to reduce these differences and consider the impact of the human body, nonlinear (Soft-computing) adaptive methods were used as described in the following section.

3.2. Experimental results: testing adaptive neuro-fuzzy inference systems

To evaluate the effectiveness of the ANFIS-based fECG filtering approach, we devised experiments using ANFISs with hybrid learning algorithms. These systems were tested with uniquely synthesized data, which comport well with real data acquired from clinical practice. Different ANFIS architectures were implemented. These structures were labeled as ANFIS1 to ANFIS5 and their parameters are summarized in **Table 2**.

	ANFIS structure				
	ANFIS 1	**ANFIS 2**	**ANFIS 3**	**ANFIS 4**	**ANFIS 5**
TNN	21	53	101	165	245
NLP	12	48	108	192	300
NNP	12	24	36	48	60
TNP	24	72	144	240	360
NFR	4	16	36	64	100

TNN—Total Number of Nodes, NLP—Number of Linear Parameters, NNP—Number of Nonlinear Parameters, TNP—Total Number of Parameters, NFR—Number of Fuzzy Rules.

Table 2. Details of ANFIS structures used in our experiments.

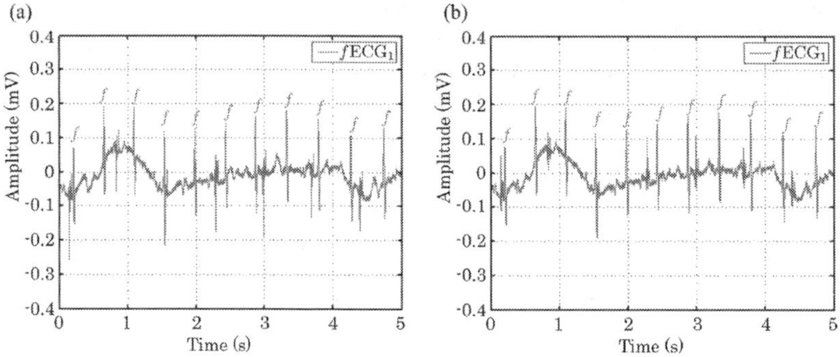

Figure 13. Output waveforms—results of filtering aECG signals (a) using ANFIS1 and (b) using ANFIS2.

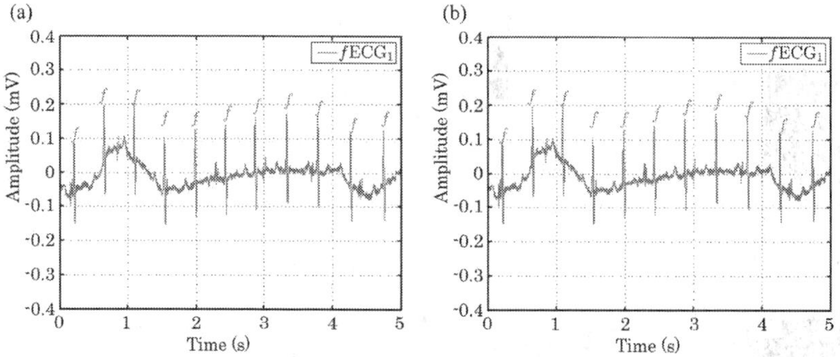

Figure 14. Output waveforms—results of filtering aECG signals (a) using ANFIS3 and (b) using ANFIS4.

Figures 13 and **14**, display the output waveforms (filtering results) for each ANFIS structure used in the experiments. **Figure 13a** shows results of filtering aECG signals using ANFIS1, (**Figure 13b**) ANFIS2.

Figure 14a shows results of filtering aECG signals using ANFIS3, (**Figure 14b**) ANFIS4. The ANFIS systems provide better results than conventional adaptive algorithms (LMS, NLMS, RLS, FTF) in time and frequency (**Figures 15** and **16**) domain.

Tables 2, **3** shows that different ANFIS structures produced different filtering results as measured by the performance metrics. The results of more complex ANFIS structures were almost identical and are not presented here. We observe that by increasing the complexity of the ANFIS structure, its computing power grows disproportionately but the improvement in filtering quality is not that significant. This fact helps us decide not to consider more complex ANFIS structures for online filtering. With this in mind, ANFIS3 seems to be the most appropriate structure for online filtering and produces acceptable results.

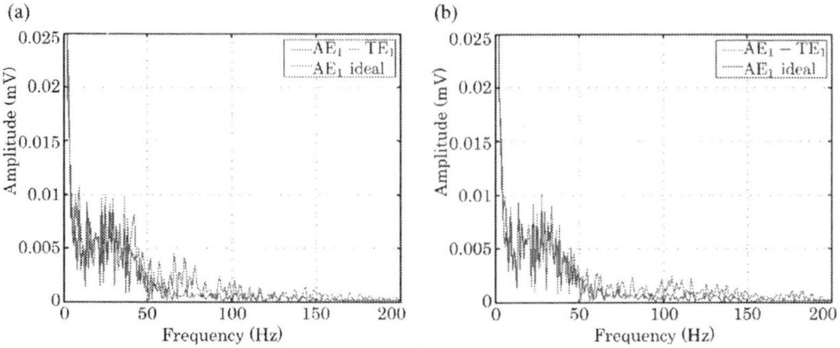

Figure 15. Amplitude spectrum—results of filtering aECG signals (a) using ANFIS1 and (b) using ANFIS2.

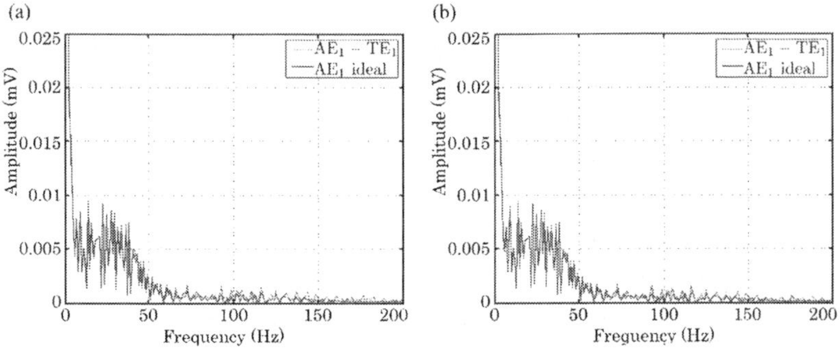

Figure 16. Amplitude spectrum—results of filtering aECG signals (a) using ANFIS3 and (b) using ANFIS4.

Figures 15 and **16** show the amplitude spectrums illustrating the filtering results achieved by ANFISs. **Figure 15a** shows the results for ANFIS1 and **Figure 15b** for ANFIS2. Each graphs illustrates the ideal fECG amplitude spectrum (Ideal AE1) together with the amplitude spectrum of the output signal from adaptive system used for each ANFIS structure.

Figure 16a shows the results for ANFIS3 and **Figure 16b** for ANFIS4. The results are summarized in **Table 3**.

Figure 17 shows the relationship between SNRin and SNRout for ANFIS1-ANFIS5. Over 100 independent experiments were performed to obtain the results reported here. We observe that these systems achieve very similar results but the processing time increases significantly with the growing complexity of their architectures.

The main advantage of the results reported here, which distinguishes them from many findings reported in the literature, is that these are achieved and tested by using clinical-quality synthetic data (identical to the real data generated by the underlying nonlinear physiological systems in the human body), thanks to the in-built capabilities of our unique signal generator.

This ensures objective assessment of the fECG signal separation quality based on quantitative performance measures (SNR, RMSE, SNRin and SNRout).

Name of adaptive filter used		Electrodes combination	SNR_{in} (dB)	SNR_{out} (dB)	SNR_{imp} (dB)	RMSE (–)	Processing time (min:s)
Adaptive neuro-fuzzy inference system	ANFIS 1	$TE_1 \leftrightarrow AE_1$	−16.0036	−7.5799	8.4237	0.1380	01:53
		$TE_2 \leftrightarrow AE_2$	−33.0132	−16.1208	16.8924	0.2723	
		$TE_2 \leftrightarrow AE_3$	−27.7029	−13.7670	13.9359	0.2190	
		$TE_1 \leftrightarrow AE_4$	−23.3521	−12.6453	10.7068	0.1526	
	ANFIS 2	$TE_1 \leftrightarrow AE_1$	−16.0036	−3.0966	12.9070	0.0595	03:33
		$TE_2 \leftrightarrow AE_2$	−33.0132	−7.9803	25.0329	0.1479	
		$TE_2 \leftrightarrow AE_3$	−27.7029	−8.5094	19.1935	0.1361	
		$TE_1 \leftrightarrow AE_4$	−23.3521	−6.1662	17.1859	0.0989	
	ANFIS 3	$TE_1 \leftrightarrow AE_1$	−16.0036	−1.0318	14.9718	0.0046	06:31
		$TE_2 \leftrightarrow AE_2$	−33.0132	−3.7510	29.2622	0.0562	
		$TE_2 \leftrightarrow AE_3$	−27.7029	−2.4146	25.2883	0.0533	
		$TE_1 \leftrightarrow AE_4$	−23.3521	−1.1298	22.2223	0.0384	
	ANFIS 4	$TE_1 \leftrightarrow AE_1$	−16.0036	−0.3004	15.7032	0.0043	12:48
		$TE_2 \leftrightarrow AE_2$	−33.0132	−3.6457	29.3675	0.0554	
		$TE_2 \leftrightarrow AE_3$	−27.7029	−2.3999	25.3030	0.0582	
		$TE_1 \leftrightarrow AE_4$	−23.3521	−1.1056	22.2465	0.0381	
	ANFIS 5	$TE_1 \leftrightarrow AE_1$	−16.0036	−0.3009	15.7027	0.0043	18:52
		$TE_2 \leftrightarrow AE_2$	−33.0132	−3.5974	29.4158	0.0590	
		$TE_2 \leftrightarrow AE_3$	−27.7029	−2.4017	25.3012	0.0453	
		$TE_1 \leftrightarrow AE_4$	−23.3521	−1.0327	22.3194	0.0373	

Table 3. Experimental results for ANFIS 1—ANFIS 5 using synthetic fECG and mECG signals generated by our novel simulator.

We should emphasis that as the ECG signals in their path from the thorax to the abdominal electrodes experience nonlinearities, the linear adaptive algorithms that are only suitable for the fHR estimation are insufficient to capture their necessary morphological details to facilitate accurate ST segment analysis (STAN). We observed that the morphological differences between the original (ideal) and recovered fECG signals by using these linear algorithms are so significant that the recovered signals cannot be used to reliably detect fetal hypoxic conditions. In other words, when clinical-quality synthetic data such as those produced by our software-controlled generator were used, adaptive algorithms were unable to adequately suppress the undesirable signals, particularly the mECG signals. For this reason, besides the

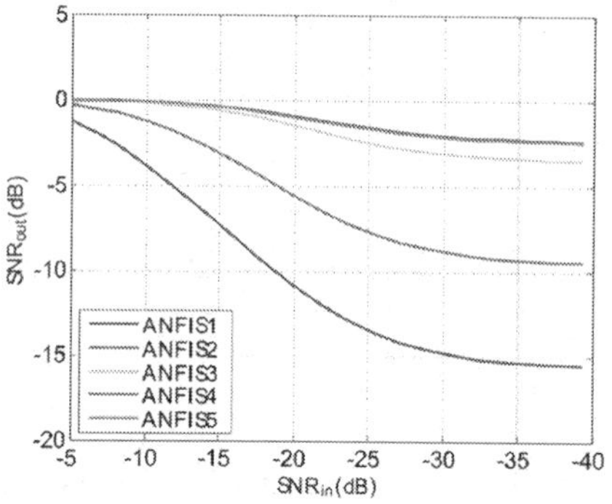

Figure 17. The Relationship between SNRin and SNRout for ANFIS31-ANFIS5.

classical adaptive approaches, soft computing techniques had to be used in our experiments to produce acceptable outcomes.

Comparing the results of ANFIS (**Table 1**) and adaptive algorithms (**Table 3**) for fECG extraction, it is demonstrated that ANFIS produces better results based upon SNR and RMSE improvements. A general disadvantage of using ANFISs is their longer processing time during network training, especially in the more sophisticated networks, which are necessary for more complex systems. From our experimental results in (**Table 1**), it is evident that the final values of SNRouts are closely related to SNRins.

4. Conclusion

In this chapter we focused on the use of advanced adaptive signal processing methods and highlighted the advantage of nonlinear methods such as adaptive neuro-fuzzy inference systems in enabling researchers to develop more reliable and accurate approaches in extracting diagnostic quality fECG signals and consequently facilitate the accurate detection of hypoxic conditions during pregnancy and labor. We included recent research findings from the most relevant engineering and medical literature and added our own contributions to the field. It is important to emphasize that currently only a fraction of the vast amount of diagnostic information in the abdominal ECG is used in clinical practice. Therefore, maximizing information extraction from fECG and CTG signals for the timely and reliable detection of fetal hypoxia is of tremendous clinical interest. This is a major challenge in signal processing and modern obstetrics as accurate determination of the fetus' status during pregnancy and labor is highly dependent on the quality of abdominal ECG monitoring, fECG signal filtering, and the consequent analysis of CTG and ST segments. In this chapter we also used a novel multichannel adaptive system that was designed, implemented, and validated by the authors to generate clinical-quality data and test and compare a variety of relevant signal processing algorithms. The primary component of this system is its adaptive block and the associated back-propagated mechanism that requires two inputs for each channel: the desired and the actual.

Our experimental results using clinical-quality synthetic data generated by our novel signal generator revealed that it offers the potential to significantly refine the diagnostic quality of the noninvasive aECG signals. We could safely conclude that following the approach presented in this chapter researchers and clinicians could acquire high quality fetal heartbeat and uterine contraction data and extract clinically significant features for reliable and accurate detection of hypoxic conditions in the fetus. As such, we are hopeful that our contribution here facilitates the development and advancement of new diagnostic methods based on transabdominal CTG + STAN. We envision that the future of fetal monitoring will greatly benefit from sophisticated diagnostic instrumentation equipped with state-of-the-art transabdominal CTG and STAN capabilities.

Author details

Radek Martinek[1*], Radana Kahankova[1], Hana Skukova[1], Jaromir Konecny[1], Petr Bilik[1], Jan Zidek[1] and Homer Nazeran[2]

*Address all correspondence to: radek.martinek@vsb.cz

1 Department of Cybernetics and Biomedical Engineering, VSB-Technical University of Ostrava, Ostrava-Poruba, Czech Republic

2 Department of Electrical and Computer Engineering, University of Texas El Paso, El Paso, TX, USA

References

[1] Amer-Wahlin I., Hellsten Ch., Noren H., Hagberg H., Herbst A., Kjellmer Ingemar, Lilja h., Lindoff C., Mansson M., Martensson L., others. Cardiotocography only versus cardiotocography plus ST analysis of fetal electrocardiogram for intrapartum fetal monitoring: a Swedish randomised controlled trial. The Lancet. 2001;358(9281):534–538.

[2] Freeman R. K., Garite T. J., Nageotte M. P., Miller L. A. Fetal heart rate monitoring. Lippincott Williams & Wilkins, Philadelphia; 2012.

[3] Soothill P. W., Nicolaides K. H., Campbell S. Prenatal asphyxia hyperlacticaemia hypoglycaemia and erythroblastosis in growth retarded fetuses. BMJ. 1987;294(6579): 1051–1053.

[4] Jagannath D. J., Selvakumar A. I. Issues and research on foetal electrocardiogram signal elicitation. Biomedical Signal Processing and Control. 2014;10(1):224–244. doi:10.1016/j.bspc.2013.11.001

[5] Burattini L., Agostinelli A., Grillo M., Biagini A., Giuliani C., Burattini L., Fioretti S., Di Nardo F., Giannubilo S. R., Ciavattini A. Noninvasive fetal electrocardiography: an overview of the signal electrophysiological meaning, recording procedures, and processing techniques. Annals of Noninvasive Electrocardiology. 2015;20(4):303–313. doi:10.1111/anec.12259

[6] Neoventa [Internet]. http://www.neoventa.com/products/stan. Accessed 2015-05-31

[7] Martinek R., Kelnar M., Vojcinak P., Koudelka P., Vanus J, Bilik P., Janku P, Nazeran H., Zidek J. Virtual simulator for the generation of patho-physiological foetal ECGs during the prenatal period. Electronics Letters. 2015;51(22):1738–1740.

[8] Martinek R., Kelnar M., Koudelka P., Vanus J., Bilik P., Janku P., Nazeran H., Zidek J. A novel LabVIEW-based multi-channel non-invasive abdominal maternal-fetal electrocardiogram signal generator. Physiological Measurement. 2016;37(2):238.

[9] Datian Ye, Xuemei Ouyang. Application of wavelet analysis in detection of fetal ECG. In: Engineering in Medicine and Biology Society, 1996. Bridging Disciplines for Biomedicine. Proceedings of the 18th Annual International Conference of the IEEE; 1996. p. 1043–1044.

[10] Khamene Ali, Negahdaripour Shahriar. A new method for the extraction of fetal ECG from the composite abdominal signal. IEEE Transactions on Biomedical Engineering. 2000;47(4):507–516.

[11] Karvounis E. C., Papaloukas C., Fotiadis Dimitrios I., Michalis L. K. Fetal heart rate extraction from composite maternal ECG using complex continuous wavelet transform. In: Computers in Cardiology, 2004; 2004. p. 737–740.

[12] Shi Zhenwei, Zhang Changshui. Semi-blind source extraction for fetal electrocardiogram extraction by combining non-Gaussianity and time-correlation. Neurocomputing. 2007;70(7):1574–1581.

[13] Bergveld Piet, Meijer Wietze JH. A new technique for the suppression of the MECG. IEEE Transactions on Biomedical Engineering. 1981;(4):348–354.

[14] Al-Zaben A, Al-Smadi A. Extraction of foetal ECG by combination of singular value decomposition and neuro-fuzzy inference system. Physics in Medicine and Biology. 2005;51(1).

[15] De Lathauwer Lieven, De Moor Bart, Vandewalle Joos. Fetal electrocardiogram extraction by blind source subspace separation. IEEE Transactions on Biomedical Engineering. 2000;47(5):567–572.

[16] Sameni R., Jutten Ch., Shamsollahi H. B. What ICA provides for ECG processing: application to noninvasive fetal ECG extraction. In: IEEE International Symposium on Signal Processing and Information Technology, 2006; 2006. p. 656–661.

[17] Marossero D. E., Erdogmus D., Euliano N., Principe J. C., Hild K. E. Independent components analysis for fetal electrocardiogram extraction: a case for the data efficient Mermaid algorithm. In: 2003 IEEE 13th Workshop on Neural Networks for Signal Processing, 2003. NNSP'03; 2003. p. 399–408. doi:10.1109/NNSP.2003.1318039.

[18] Hon E. H., Lee S. T. Averaging techniques in fetal electrocardiography. Medical Electronics and Biological Engineering. 1964;2(1):71–76.

[19] Outram NJ, Ifeachor EC, Van Eetvelt PWJ, Curnow JSH. Techniques for optimal enhancement, feature extraction of fetal electrocardiogram. In: IEE Proceedings Science, Measurement and Technology; 1995. p. 482–489.

[20] Favret A. G. Computer matched filter location of fetalR-waves. Medical and Biological Engineering. 1968;6(5):467–475.

[21] Widrow B., Glover Jr J. R., McCool J. M., Kaunitz J., Williams Ch. S., Hearn R. H., Zeidler J. R., Dong Jr E., Goodlin R. C. Adaptive noise cancelling: principles and applications. Proceedings of the IEEE. 1975;63(12):1692–1716.

[22] Martinek R., Kelnar M., Koudelka P., Vanus J, Bilik P., Janku P., Nazeran H., Zidek J. Enhanced processing and analysis of multi-channel noninvasive abdominal foetal ECG signals during labor and delivery. Electronics Letters. 2015;51(22):1744–1746.

[23] Martinek R, Zidek J. Refining the diagnostic quality of the abdominal fetal electrocardiogram using the techniques of artificial intelligence. Przeglad Elektrotechniczny. 2012;88(12 B):155–160.

[24] Swarnalatha R, Prasad DV. A novel technique for extraction of FECG using multi stage adaptive filtering. Journal of Applied Sciences. 2010;10(4):319–324.

[25] Wu Shuicai, Shen Yanni, Zhou Zhuhuang, Lin Lan, Zeng Yanjun, Gao Xiaofeng. Research of fetal ECG extraction using wavelet analysis and adaptive filtering. Computers in Biology and Medicine. 2013;43(10):1622–1627.

[26] Poularikas Alexander D, Ramadan Zayed M. Adaptive filtering primer with MATLAB. CRC Press, Boca Raton; 2006.

[27] Wei Zheng, Xueyun Wei, Hongxing Liu, others. Noninvasive fetal ECG estimation using adaptive comb filter. Computer Methods and Programs in Biomedicine. 2013;112(1).

[28] Shadaydeh Maha, Xiao Yegui, Kriedieh Ward Rabab. Extraction of fetal ECG using adaptive Volterra filters. In: 16th European Signal Processing Conference, 2008; 2008. p. 1–5.

[29] Sameni R., Shamsollahi M. B., Jutten Ch., Babaie-Zadeh M. Filtering noisy ECG signals using the extended Kalman filter based on a modified dynamic ECG model. In: Computers in Cardiology, 2005; 2005. p. 1017–1020.

[30] Niknazar M., Rivet B., Jutten Ch. Fetal ECG extraction by extended state Kalman filtering based on single-channel recordings. IEEE Transactions on Biomedical Engineering. 2013;60(5):1345–1352.

[31] Bin Ibne Reaz Mamun, Wei Lee Sze. Adaptive linear neural network filter for fetal ECG extraction. In: Proceedings of International Conference on Intelligent Sensing and Information Processing, 2004; 2004. p. 321–324.

[32] Sameni Reza. Extraction of fetal cardiac signals from an array of maternal abdominal recordings [dissertation]. Sharif University of Technology, Tehran, Iran; 2008.

[33] Alexander Thomas S. Adaptive signal processing: theory and applications. Springer Science & Business Media, New York; 2012.

[34] National instruments. Least mean square (LMS) Adaptive filter–developer zone—national instruments. [Internet]. 10. June 2009. Accessed 26-2-2016

[35] Akay Metin. Biomedical signal processing. Academic Press, Cambridge; 2012.

[36] Cioffi J. M., Kailath T. Fast, recursive-least-squares transversal filters for adaptive filtering. IEEE Transactions on Acoustics, Speech and Signal Processing. 1984;32(2): 304–337.

[37] Martinez M., Calpe J., Soria E., Guerrero J. F., Camps G., Gomez L. Methods to evaluate the performance of fetal electrocardiogram extraction algorithms. In: Computers in Cardiology 2001; 2001. p. 253–256.

[38] Mahfuz Ejaz, Wang Chunyan, Ahmad M Omair. A high-throughput DLMS adaptive algorithm. In: IEEE International Symposium on Circuits and Systems, 2005. ISCAS 2005; 2005. p. 3753–3756.

[39] Suja Priyadharsini S, Edward Rajan S, Saranya S. An efficient soft-computing technique for extracting fetal. In: Programmable Device Circuits and Systems; 2011. p. 1–7.

[40] Nasiri Maryam. Fetal electrocardiogram signal extraction by ANFIS trained with PSO method. International Journal of Electrical and Computer Engineering. 2012;2:247 (IAES Institute of Advanced Engineering and Science).

[41] Martinek R., Sincl A., Vanus J., Kelnar M., Bilik P., Machacek Z., Zidek J. Modelling of fetal hypoxic conditions based on virtual instrumentation. In: Proceedings of the Second International Afro-European Conference for Industrial Advancement AECIA 2015; 2016. p. 249–259.

[42] Martinek R., Zidek J. A system for improving the diagnostic quality of fetal electrocardiogram. Przeglad Elektrotchniczny (Electrical Review), R. 2012;88:164–173.

[43] Caydas U., Hascalik A., Ekici S. An adaptive neuro-fuzzy inference system (ANFIS) model for wire-EDM. Expert Systems with Applications. 2009;36(3):6135–6139.

[44] Jang Jyh-Shing Roger. ANFIS: adaptive-network-based fuzzy inference system. IEEE Transactions on Systems Man and Cybernetics. 1993;23(3):665–685.

[45] Swarnalatha R, Prasad DV. Maternal ECG cancellation in abdominal signal using ANFIS and wavelets. Applied Science. 2010;10:868–877.

[46] Martinek R. The Use of complex adaptive methods of signal processingfor refining the diagnostic quality of the [dissertation]. VSB—Technical University of Ostrava; 2014.

[47] Nguyen Hung T, Prasad Nadipuram R, Walker Carol L, Walker Elbert A. A first course in fuzzy and neural control. CRC press, Boca Raton; 2002.

[48] Wang Peizhuang, Tan Shaohua. Soft computing and fuzzy logic. Soft Computing. 1997;1(1):35–41.

[49] Zadeh Lotfi A. Soft computing and fuzzy logic. IEEE Software. 1994;11(6):48.

[50] Zadeh Lotfi A. Fuzzy logic=computing with words. IEEE Transactions on Fuzzy Systems. 1966;4(2):103–111.

[51] Zadeh Lofti A. Fuzzy logic, neural networks, and soft computing. Communications of the ACM. 1994;37(3):77–85.

[52] Camps-Valls Gustavo, Martinez-Sober Marcelino, Soria-Olivas Emilio, Magdalena-Benedito Rafael, Calpe-Maravilla Javier, Guerrero-Martinez Juan. Foetal ECG recovery using dynamic neural networks. Artificial Intelligence in Medicine. 2004;31(3):197–209.

[53] Kedir-Talha M, Guettouche MA, Bousbia-Salah A. Combination of a FIR filter with a genetic algorithm for the extraction of a fetal ECG. In: Proceedings of International Conference on Circuit, Systems and Signals (CSS10), Malta, September 2010, p. 76; 2010.

[54] Kam Amit, Cohen Arnon. Detection of fetal ECG with IIR adaptive filtering and genetic algorithms. In: 1999 IEEE International Conference on Acoustics, Speech, and Signal Processing, 1999; 1999. p. 2335–2338.

[55] Kezi Selva Vijila CP, Kanagasabapathy S. Johnson. Adaptive neuro fuzzy inference system for extraction of fecg. In: Annual IEEE India Conference—Indicon; 2005. p. 224–227. doi:10.1109/INDCON.2005.1590160.

[56] Assaleh Khaled. Extraction of fetal electrocardiogram using adaptive neuro-fuzzy inference systems. IEEE Transactions on Biomedical Engineering. 2007;54(1):59–68.

[57] Jalaleddine Sateh, Hutchens Chriswell G, Strattan Robert D, Coberly William A. ECG data compression techniques-a unified approach. IEEE Transactions on Biomedical Engineering. 1990;37(4):329–343.

[58] Wen He, Meng Zhuo, Teng Zhaosheng, Guo Siyu, Yang Yuxiang. Comparative study of influence of noise on power frequency estimation of sine wave using interpolation FFT. Fluctuation and Noise Letters. 2014;13(03)

[59] Wen He, Teng Zhaosheng, Guo Siyu. Triangular self-convolution window with desirable sidelobe behaviors for harmonic analysis of power system. IEEE Transactions on Instrumentation and Measurement. 2010;59(3):543–552.

[60] Martinek R., Kelnar M., Vanus J., Bilik P., Zidek J. A robust approach for acoustic noise suppression in speech using ANFIS. Journal of Electrical Engineering. 2015;66(6):301–310.

[61] Kromer Pavel, Prauzek Michal, Musilek Petr. Harvesting-aware control of wireless sensor nodes using fuzzy logic and differential evolution. In: 2014 Eleventh Annual IEEE International Conference on Sensing, Communication, and Networking Workshops (SECON Workshops); 2014. p. 51–56.

[62] Martinek R., Zidek J. Use of adaptive filtering for noise reduction in communications systems. In: 2010 International Conference on Applied Electronics (AE). 2010; p. 1–6.

[63] Martinek R., Zidek J. The real implementation of ANFIS channel equalizer on the system of software-defined radio. IETE Journal of Research. 2014;60(2):183–193.

Chapter 4

Network Theoretical Approach to Describe Epileptic Processes

Ancor Sanz-García, Rafael G. de Sola,
Lorena Vega-Zelaya, Jesús Pastor and
Guillermo J. Ortega

Additional information is available at the end of the chapter

http://dx.doi.org/10.5772/63914

Abstract

Epilepsy is characterized by recurrent unprovoked seizures. Recent studies suggest that seizure generation may be caused by the abnormal activity of the entire network. This new paradigm requires new tools and methods for its study. In this sense, synchronization by linear as well as nonlinear measures are used to determine network structure and functional connectivity of neurophysiological data. Electroencephalography (EEG) data can be analyzed using each electrode's activity as a node of the underlying cortical network. The information provided by the synchronization matrix is the basic brick upon which several lines of analysis can be performed thereafter. Detection of community structures, identification of centrality nodes, transformation of the underlying network into a simpler one, and the identification of the basic network architecture are only some of the many lines of basic works that can be done in order to characterize the epilepsy as a network disease. This chapter describes new approaches in network epilepsy, provides mathematical concepts in order to understand the complex network analyses, and reviews the advances in network analyses and its application to epilepsy research.

Keywords: synchronization, temporal lobe epilepsy, electroencephalography, electrocorticography, graph theory, centrality measures, community structures, small-world

1. Introduction

Epilepsy is one of the most common neurological disorders characterized by recurrent unprovoked seizures. It has an incidence of 50–100 cases per 100,000 persons/year in developed countries, and about 50 million people around the world suffer from epilepsy [1]. The most common type of epilepsy is temporal lobe epilepsy (TLE). Unfortunately, a high percentage of TLE patients are resistant to drug treatment presenting the so-called drug-resistant epilepsy (DRE). Surgery is recommended in such cases as the only curative/palliative alternative. Meta-analyses from literature indicate that 66% of the patients are seizure free in the first two years following the surgery [2]. However, seizures persist in one third of the operated patients; thus, an accurate localization of the epileptogenic zone (EZ) is vital for a good surgery outcome.

Traditionally, seizures in partial epilepsy have been considered to arise from a single focus that recruits other regions in order to spread. Very recently however, an alternative point of view has suggested that ictogenesis (seizure generation) may be caused by the abnormal activity of the entire network instead of being provoked by pathological isolated areas, such as the EZ [3–7]. In fact, the Commission on Classification and Terminology from the International League Against Epilepsy (ILAE) has proposed a new approximation for the classification of seizures and epilepsy types [8]. The new recommended terminology redefines focal seizures as those originating at some points within networks spatially limited to only one hemisphere, or bilaterally distributed in the case of generalized seizures. This new perspective has been mostly inspired, on one hand from the network analyses of data obtained from several diagnostic tools that are routinely used nowadays in the pre-surgical evaluation of epilepsy, such as functional magnetic resonance imaging (MRI) (for review see [9]) or even from the classical one, the electroencephalography (EEG) [4,6,9]. On the other hand, the development of the connectome concept, which describes the connectivity among different brain areas in physiological conditions, has boosted this new perspective. In this regard, three types of connectivity have been defined: the anatomical, which are the structural connections linking the neurons or brain regions; the functional, which describes the statistical relations between activity of different neurons or brain regions; and lastly, the effective connectivity, which assesses the causal effects of one region or neuron to another [10].

Similarly to the impact of network analyses in seizure definition and classification, this new perspective on epilepsy has also modified the traditional pre-surgical evaluation of TLE patients. The traditional zone-oriented approach [11] has been complemented with the recent advent of the network perspective [4,5]. The zone-oriented concept of the EZ can be interpreted under this new point of view as a key property, yet unknown, of the network in charge of generating and sustaining the seizures. The new aim in epilepsy surgery would be now targeted at destroying this particular property of the network instead of being aimed at resecting a cortical area such as the EZ, as it is traditionally done. The importance of identification, delimitation and characterization of the epileptic network clearly shows up when the above considerations are taking into account and justify the great number of works devoted to this issue [12–18].

In this regard, EEG data obtained either invasively or non-invasively, can be now analyzed using each electrode's activity as a node of the underlying cortical network. In this framework, it is fashionable to estimate functional connectivity between cortical areas covered by neurophysiological electrodes through the correlation or synchronization – both terms are used interchangeably through the chapter, although their meanings are not exactly the same

– matrix of the electrodes' time series. The information provided by the synchronization matrix is the basic brick upon which several lines of research can be built. One of them is the study of hierarchical organization, which through the construction of hierarchical trees provides insightful information of closeness and farness of node's neurophysiological activity. The application of this technique characterizes the underlying dynamical behavior of the whole network and allows to easily reveal those regions of highly synchronized nodes, i.e. synchronization clusters. Also, by combining the information provided by the synchronization matrices, and several techniques borrowed from graph theory, give us the opportunity to study the underlying cortical network as a whole entity.

It is important to emphasize that the network perspective is not only limited to EEG studies but functional MRI (fMRI) also provides evidences of this emerging view. Despite fMRI possesses a lower temporal resolution and is an indirect measure of neuronal activity, as compared with EEG, fMRI has a good spatial resolution and obviously provides critical data about the functional connectivity of the whole brain. In this regard, simultaneous fMRI and EEG studies have revealed the validity of the BOLD analysis [19], showing that changes in the epileptic network actually occur (for a review see [9]). In line with fMRI and EEG data, pathological and structural studies also corroborate the study of epilepsy as a network disorder, since histological studies have revealed neuronal cell loss, gliosis, axonal sprouting, among others in hippocampus of the TLE patients [20]. Indeed, similar changes have been described in adjacent areas to the hippocampus, as the amygdala, the entorhinal cortex, the parahippocampal cortex, certain areas of the temporal cortex, and even other cortical and thalamic areas [9]). All these results suggest that network reorganization occurs in epilepsy from a histological point of view. Similarly, anatomical and diffusion MRI studies supported the histological results, showing loss of grey matter and disorganization of fibers.

In this chapter, we review in a first step, the synchronization concept, its different types and estimators in order to provide an overview of the basic element needed to perform a network analysis of neurophysiological data. Secondly, we present the new network epilepsy approach with detailed and comprehensive explanations of the underlying mathematical concepts needed to fully understand the involved concepts and methodologies. Our aim, thus, is to present the recent advances in network analyses, covering issues such as hierarchical classifications, community detection, centrality measures identification and topological organization in order to state their outstanding relevance in a particular field of biomedical signal processing as it is the evaluation of pre-surgical neurophysiological recordings coming from DRE patients.

2. Synchronization

2.1. Concepts

Synchronization is the building block upon which functional networks are constructed. Once a functional relation between any two cortical areas is established, for all of the recording electrodes, the first step in the network construction can be confidently performed. This is the main reason why synchronization estimation is so critical. In recent years there has been a rising number of works dealing with this issue, and in particular in their application to brain functional connectivity. It is important to emphasize that not a single method can accurately provide the true, if something like this exists, underlying synchronization, because each one possesses its drawbacks and benefits. In any case, several methods should be employed.

Synchronization is closely related to epilepsy, as commented Section 1. Conceptually, synchronization is the adjustment of the internal rhythms of two systems due to an existing (weak) interaction between them [21]. Furthermore, it is essential for synchronization to be established, the presence of two or more self-sustained oscillators with their own rhythms.

2.2. Types of synchronization

The most intuitive form of synchronization is the *frequency synchronization*. That is, two systems, e.g. x and y, through mutual interaction adjust their own rhythms to share a similar frequency $\omega_x = \omega_y$. Frequency synchronization does not need that exactly the same frequency be shared by two synchronized systems, however. Instead, the following relation is valid: $n_x\omega_x - n_y\omega_y = 0$, where n_x and n_y are integers. Moreover, in chaotic systems it seems to be less restrictive than other types of synchronization (phase or identical), because instantaneous values of variables may be different.

In addition to frequency synchronization, phase can also become synchronized, this occurs when the phases of both systems are similar. Similar to the frequency synchronization, both phases do not have to be necessarily equal since *phase synchronization* also occurs when a constant time difference exists between them. The synchronization of both frequency and phase is called *identical synchronization*; likewise, when a lag exists between them it is termed *lag synchronization*.

When a system, slave, is under governance of another system, master, there is a unidirectional relation. In this condition, the synchronization is called *generalized synchronization*. To describe this type of synchronization in EEG data is necessarily to use the embedding theorem which allows to reconstruct an equivalent system of the original one.

The different types of synchronization described above can be used for the analysis of synchronization of more than two systems. However, in those cases it is better to use the *full synchronization* in order to determine the synchronization of the whole oscillators' interaction rather than the interaction between each pair. The oscillator is replaced by its instantaneous phase and intrinsic synchrony, and the spatial distribution is added. The spatial distribution is formed by the coupling strength between adjacent oscillators and the existence or absence of each link. The whole system synchronization is achieved by an order parameter that provides information about the phase coherence of the oscillator and the degree of synchronization.

2.3. Synchronization estimation

Several numerical methodologies have been used to determine the synchronization; the most common are the cross-correlation [22] and the cross-spectrum analysis. However, new methodologies have gradually been introduced to evaluate networks connectivity. This is, to address the temporal correlation between spatially separated nodes or the influence of one neural system over another, i.e. the functional and effective connectivity, respectively (see Section 3).

Broadly speaking, two types of correlation exist, the linear and nonlinear correlations [23]. Both have its pros and cons, however. Usually, a tradeoff between accuracy and speed appear at the time of deciding which method to use. These two premises are critical, mainly the first one, because of the high number of comparisons, which increases the number of false positive, i.e. the number of significant correlation that does not correspond to real signification. In these cases, statistical methods have to be used in order to improve the sensitivity, for instance, the classical methodology, such as Bonferroni correction, or more sophisticated systems as the surrogate data testing.

Cross-correlation is essentially an amplitude method in the sense that it quantifies co-movements in two time series by "comparing" amplitudes in the signals. Pearson correlation coefficient [22] is the best known method for synchrony calculation by cross-correlation since it is a function of the lag between signals; the lag has to be specified. For two discretized time series, x_i and x_j, at times k, in a 0 lag condition the ρ_{ij} is calculated as follows,

$$\rho_{ij}(0) = \frac{\sum_{k=1}^{N_{wins}} (x_i(k) - \overline{x}_i)(x_j(k) - \overline{x}_j)}{\sqrt{\sum_{k=1}^{N_{wins}} (x_i(k) - \overline{x}_i)^2 \sum_{k=1}^{N_{wins}} (x_j(k) - \overline{x}_j)^2}} \tag{1}$$

Correlation coefficient range is between $-1 \leq \rho_{ij} \leq 1$, calculated in the time window N_{win}. Higher negative values mean a higher inverse linear correlation whereas higher positive values indicate a higher linear relation between time series.

A zero cross-correlation in Eq. (1) does not mean that there is no correlation because nonlinear correlation could also exist; this is because nonlinear analyses compare other parameters rather than amplitude [24]. For instance, phase synchronization measures the differences in phases. In two time series x_i and x_j, at times k, the phase synchronization is calculated by the mean phase coherence, R_{ij},

$$R_{ij} = \left| \frac{1}{N_{wins}} \sum_{k=1}^{N_{wins}} e^{\Delta\alpha_{ij}(k)} \right| \tag{2}$$

Correlation coefficient range is between $0 \leq R_{ij} \leq 1$, calculated in the time window N_{win}, where $\Delta\alpha_{ij}(k)$ is the instantaneous phase difference at the discretized time k.

Information theory [25], through mutual information, has been used to evaluate the association between the two variables. Roughly speaking, if the (Shannon) entropy of a time series x_i quantifies the degree of uncertainty about future values of it, the mutual information between x_i and y_i quantifies the reduction of uncertainty in x_i knowing y_i. Thus, mutual information between two time series evaluates the statistical association between them. Mutual information range from zero to positive values, such that zero means that the two systems are statistically independent.

Figure 1 displays the typical synchronization patterns calculated from a typical recording of a scalp EEG. In **Figure 1A**, the correlation matrix calculated through the Pearson estimate is depicted. **Figure 1B** shows a similar calculation but using Eq. (2), corresponding to the phase synchronization. On the contrary, **Figure 1C** shows the synchronization calculation carried out by using the mutual information measure. Note that in this last case diagonal elements have been eliminated because they have not upper bound.

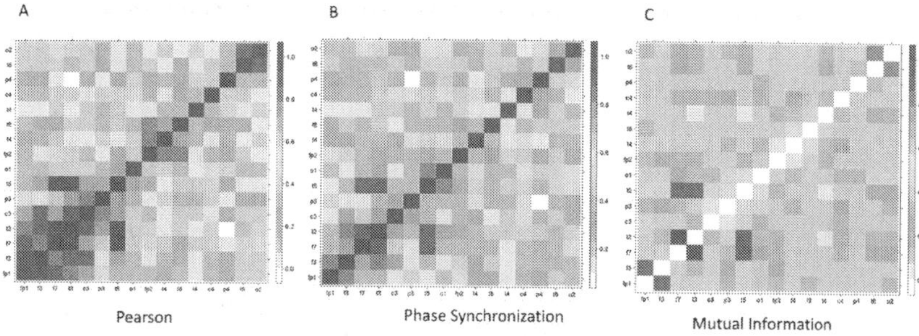

Figure 1. Synchronization patterns: synchronization matrices for three different measures calculated over 10 s of scalp recordings. (A) Pearson correlation, (B) phase synchronization using the mean phase coherence (Eq. (2)), (C) mutual information with diagonal elements set to 0 in order to make it clear the interaction between different elements (see main text).

Finally, the integration of these different synchrony measurements in a spatially extended system is what is called spatial synchronization. It is especially useful when dealing with neurophysiological data in order to gain insight of the underlying network, and becomes the main concept for several network analyses that are the subject of this chapter. The spatial synchronization is the organization of the whole network based on their nodes interaction, i.e. correlation between every pair of nodes. Firstly, transformation of the synchrony matrix to distance is required, i.e. higher correlations mean closer distances between nodes, in order to perform a hierarchical organization (see Section 3.4.1). Several methods and algorithms exist to perform the hierarchical tree; the best known is the hierarchical clustering [26]. It organizes nodes in smaller groups based on their higher correlation, or more properly their closer distance, to generate groups of highly linked nodes, known as cluster.

2.4. Synchronization and epilepsy

Late 1940s and early 1950s descriptions [27] presented epilepsy as a neurological disorder associated with excessive synchrony leading to a state of "hypersynchrony", an idea maintained recently. However, recent advances in chaos theory, complex networks and nonlinear time series analyses have reviewed this classical interpretation of synchronization and its application to seizure characterization. Nowadays, epilepsy is far from being considered as a merely hypersynchronized "state". Extent data from literature present epilepsy as a synchronization/desynchronization process, remarkably active during seizures. Dramatic changes in synchronization have been quantified during the seizure extent, highlighting thus the concept of synchronization as a truly process instead of a particular state [24,28,29]. This section will approach the relations between synchronization and epilepsy.

2.4.1. Synchronization and hypersynchronization

From a classical [30,31] and a fundamental point of view, the epileptic activity starts as a paroxysmal depolarization shift (PDS) in thousands of hypersynchronously neurons causing a mesoscopic manifestation in the EEG recordings – the well-known interictal epileptogenic discharges (IEDs) [32–34]. The spread of this activity, which in physiological conditions would remain confined due to weakened inhibition of the surrounding areas, induces the seizure start. This means that large population of neurons far from the seizure onset zone ones triggered synchronously. Synchronization thus can be studied at two different scales, i.e. local and global.

Although seizures are the pathophysiological definition of epilepsy, other EEG disturbances are considered as clinical signs. IEDs are widely accepted as an EEG marker of epilepsy since they are present in 80–90% of the patients. IEDs present different patterns in the EEG recordings, such as spikes, poly-spikes and/or sharp waves [35, 36]. The way in which PDS manifest as IEDs can be explained by local synchronization mechanisms, specifically, strong and weak mechanisms of interaction. The strong mechanism involves the direct communication between neurons by chemical or electrical (gap junctions) synapses [37]. In this regard, computational simulations of the dentate gyrus, one of the components of the hippocampal formation and an area highly involved in temporal lobe epilepsy, have demonstrated that the incorporation of a small number of highly interconnected hubs greatly increase the hyperexcitability of the network [38], modifying the network topology toward a scale-free one. Therefore, it is suggested that the strong mechanisms increase the synchronization of the whole network inducing IEDs manifestation. On the contrary, weak interactions are due to the extracellular space changes of ions concentration or electrical field transmission induced by the activity of neighboring cells [39]. Both interactions are present in physiological processes and are needed to an appropriate brain function; however, an enhancement of this processes due to the several pathological issues produces the IEDs. To summarize, IEDs can be considered as the result of a small number of neurons triggering simultaneously driven by weak and strong interactions in a process considered as hypersynchronization, or suitably, a local full synchronization process.

The clinical relevance of IEDs is reflected in the huge efforts made by the researchers to describe and identify them [40]. In fact, most of the past publications use traditional raw EEG analysis in order to lateralize and localize the seizure by the presence of IEDs, since the ipsilateral side presents a higher number of IEDs as compared to the contralateral [41–43]. The first evidence of distant synchronization in epilepsy occurred by the analyses of IEDs, demonstrating a hippocampal–entorhinal cortex interaction [44]. However when a synchronization analysis of the interictal period is performed, a critical issue has to be considered: the role of IEDs and the background interictal activity in the synchronization of interictal period. In reference [45], the authors deal with this question; they demonstrated that IEDs are a 1.2, 10, 13.3, 33.3% of the total time in scalp EEG, foramen ovale electrodes (FOE), electrocorticography (ECoG) and depth electrode recordings, respectively. These results give an idea of the contribution of IEDs in each type of recording to the synchronization measurement. Surprisingly, the intuitive association of a higher presence of IEDs in the ipsilateral side with a higher synchrony in the same side presents controversial results, with works favoring this possibility [46–48] and others not [49]. Perhaps this disagreement can be explained by the above-explained synchronization bias of the depth electrodes as compared to FOE, since works favoring higher synchronization in the ipsilateral side were performed in depth electrodes recordings.

The classical definition describes seizure as a hypersynchronization of a large amount of neurons; however, recent evidences present it as a more complex phenomena. As stated earlier, synchronization has to be considered a dynamical process and that is what happens in a seizure. An increase of desynchronization is observed in the first stages [50] followed by a large-scale increase in synchronization. Hypersynchronization has also been proposed as a seizure termination process, since excessive activation of neurons induces several mechanism of autoregulation entailing seizure termination [29]. Seizure synchronization depends on the spatial scale, the type of synchrony and the EEG patterns analyzed, among others. Despite hypersynchronization also occurs in seizure at cellular level, only a 30% of neurons change its firing rate [50] during partial seizures. This suggests that a dynamical approach is more appropriate to analyze seizure synchronization.

2.4.2. Desynchronization

The role played by desynchronization in epilepsy has been well documented in the literature [49–53]. Far from challenging the classical point of view, desynchronization has been integrated into the epilepsy facts providing a new perspective. It occurs not only at a global scale but also at the cellular level, since desynchronization has also been found in seizure-like events of cellular studies [52].

First evidences of desynchronization in seizure were reported by Gastaut et al. [54], by describing the electrodecremental seizures. Since then, several works have reported desynchronization prior to seizures onset. Desynchronization has been described in the analyses of depth electrodes recordings by using phase synchronization methodology. A decrease in the mean phase coherence occurs 15–20 min after seizure onset [53]. Similar results were found only in the beta-band [55]. Two hypotheses were derived from these results; one suggests that desynchronization favors the recruitment of regions until full synchronization is achieved [46], and the other suggests that desynchronization is the result of an adjustment process between two different recorded areas, one with physiological synchronization levels and the other with higher levels because of being part of the seizure onset zone [53]. Surprisingly, desynchronization was also described in the first half of seizures during secondary generalization [56,50]. The authors suggest that this is due to different routes and lags during the seizure propagation [57].

Not only in seizures but also during the interictal period, desynchronization also seems to play an important role in the epilepsy dynamics. Lower levels of synchronization were reported [49,58] in the ipsilateral side, as compared with the contralateral one, to seizures in TLE patients. These lower levels were maintained also during the seizures [58].

Desynchronization also occurs at the cellular level as it was demonstrated in an elegant study of Netoff and Schiff [52] in *in vitro* hippocampal slices. They show the existence of decreased synchronization immediately after ictal-like activity. They observed this change using both linear and nonlinear methods. The initial desynchronization can be explained because of the presence of microdomains of highly synchronized neuronal clusters present in ictogenesis in *in vitro* studies [59,60]. Moreover, microseizures have been described prominently in the seizure onset zone in human recordings [61–63]. Zones that present microseizures are good candidates of being part of the seizure onset. However, instead of single initial onset zone, multiple distant microdomains that synchronize and produce the macroseizure could also be another mechanism of seizure onset [29]. In both cases, the presence of different regions disturbs the synchronization estimation, that is, if synchrony is measured in a region that includes areas generating ictal activity and others remain unaffected the global results will be a desynchronization [64]. These results were also confirmed by the computational and electrophysiological models, demonstrating that seizure-like activity can be induced even when the synaptic strength of neurons is small, suggesting that higher excitatory strength is not needed to seizure induction [65].

This new perspective of epilepsy as a dynamical process has provided new insights of epilepsy and seizures, bringing more evidences that justify the complex network analyses. Although the use of synchronization measures by epileptologists is still rather marginal, a raising interest in this decade has appeared to perform analyses by using linear as well as nonlinear methods. The next section introduces concepts and methodologies of the complex network analysis, mainly based on synchronization methods, applied to neurophysiological data.

3. Networks

Traditionally the brain has been considered as a compartmentalized structure with particular functions in each area. However, over the last decade a network approach has been proposed as a more accurate model for brain functioning, leading to the appearance of the connectome concept [66], a connectivity map derived from the neurophysiological and imaging data.

Besides the network-oriented study of physiological brain connectivity, the study of several pathologies, such as epilepsy, has also been focused on the network paradigm. Evidences from

EEG studies have raised this alternative point of view. In fact, the ILAE proposed a new terminology including the network concept for the classification of seizures and epilepsies [8]. Moreover, several studies of the ictogenesis [3–6] suggested that seizures are caused by the abnormal activity of a network rather than by an isolated area malfunction. All these information, in particular the last statement, has boosted a large amount of re-analysis of EEG recordings under this new perspective [4,5,11]. Indeed, a rising number of works have appeared in the recent years supporting this idea [12–18].

To deal with this new paradigm, studies are increasingly using graph theory, a mathematical framework that studies the general properties – graphical, topological, statistical, etc. – of a set of nodes interconnected by links. Any kind of system can be described and characterized under this framework as long as the system's elements can adequately be represented by network's nodes and their relations by links. In the case of the brain, and in particular with respect to the neurophysiological characterization, network nodes are represented by cortical areas covered by the electrodes' contacts and the links between them by the existing synchronization level in their electrical activity. As shown below, the type of links considered between the cortical areas determines the type of network.

Considering the brain as a connectome, three main categories of parameters can be used to extract valuable information of the network activity. These three categories correspond to the three scales where a network can be evaluated, the most basic, centrality or nodal measures; the intermediary, community structures; and the most global, the topological organization. Centrality measures study the significance of the nodes inside the network by determining its characteristics; community structures are based on the recognition or assessment of communities of nodes, so-called clusters or modules; and topological organization uses the average of some of the centrality parameters to determine the properties and architectures of a given network. These variables can be correlated with behavioral data (task performing results), clinical data (treatment or surgery outcome) or different pathologies providing a valuable knowledge or a key target in the pathological network. Thus, in the next section, network parameters have been explained providing definitions and mathematical notations, and its use in epilepsy has been discussed.

A final remark regarding the nature of the network approach is in order here. When dealing with very large networks, like networks of networks, composed of a very large number of nodes, the proper approach to network characterization is by the statistical mechanics of network. In this sense, network characterization is preferably done by the statistical properties of their nodes and/or links. The most common examples are for instance to look at the statistical distribution of the nodes connectivity ("degree", see below) and ask whether these distributions fit in any of the standard distributions: random networks, scale-free networks, small-world networks, etc. However, these kinds of networks have typically a huge quantity of nodes, which allows to study their properties across several scales. A typical example of these kinds of networks is the internet, composed of routers and computers as nodes, and the wires and cables, which connect them as the physical links. The kind of networks we are able to characterize in studying neurophysiological records is nonetheless of a very different kind.

http://dx.doi.org/10.5772/63914

3.1. Types of networks

We briefly summarize here some basic types of networks. The most basic representation of a network is by considering only the nodes and links the network is made of. **Figure 2A** represents a network of this type composed of 34 nodes and 77 links between some of the nodes.

A step into a more complete description of a network is by adding "weights" to the links, that is, it is not only relevant whether or not two different nodes are connected by a link, but also the intensity of the links does also matter. This is represented in **Figure 2B** in such a way that links' weights are represented by lines width, such that more intense connections are represented by thicker lines.

In the case we are working on, this is represented by the estimation of the synchronization level between the electrical activities covered by the corresponding contacts.

More complex representations may be achieved when directionally is introduced in the interaction between two different nodes, as it is represented in **Figure 2C**.

Other situation including finer details is represented in **Figure 2D** such that nodes' characteristics are also included in the description, represented in this case by the nodes' sizes. Other, more complex situations can be represented in the network descriptions.

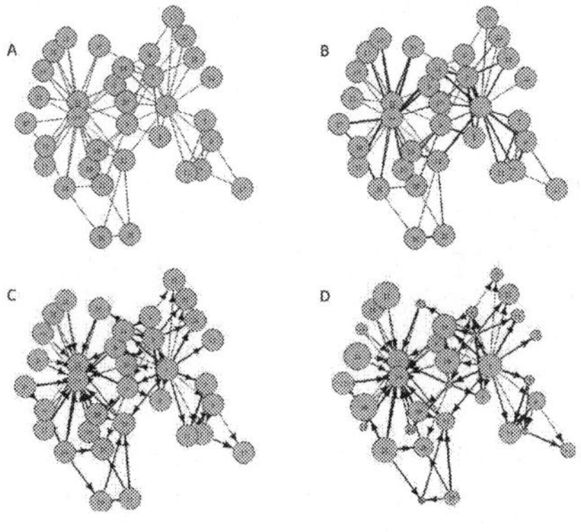

Figure 2. Different types of networks constructed with 34 nodes: (A) unweighted undirected network with equally important nodes; (B) weighted undirected network with equally important nodes; (C) weighted directed nodes with equally important nodes and (D) weighted directed network with unequally important nodes.

From a mathematical point of view, a "simple" network, as the one of **Figure 2A**, is fully characterized by giving the "adjacency matrix" whose elements are binary numbers a_{ij} denoting the presence or absence of edges, or links, between nodes $i$and j, that is $a_{ij} = 1$ if a link between node i and node j does exist and $a_{ij} = 0$ if not. Because the adjacency matrix is a symmetric matrix, that is, if a link between nodes i and j exists, then $a_{ij} = a_{ji} = 1$, the total number of links in network with adjacency matrix a_{ij} is the sum of 1's in a_{ij} divided by 2 (because of symmetry) minus the number of nodes in the networks (diagonal elements).

When dealing with networks like the one depicted in **Figure 2B**, the adjacency matrix contains the weights, that is, the 1's in the adjacency matrix are replaced by the corresponding weights w_{ij}. These kinds of networks are called weighted networks. Both kinds of networks, weighted and unweighted, are typical examples of functional connectivity representation in the neurophysiological realm. Effective connectivity on the contrary, when directionality counts are typically represented with the networks, is depicted in **Figure 2C** and **D** [10].

3.2. Networks from time series

Having described the basic properties of networks, the following step is to know how to construct the network from the recording time series. This is a critical step when the objective is to analyze neurophysiological recordings under a network perspective.

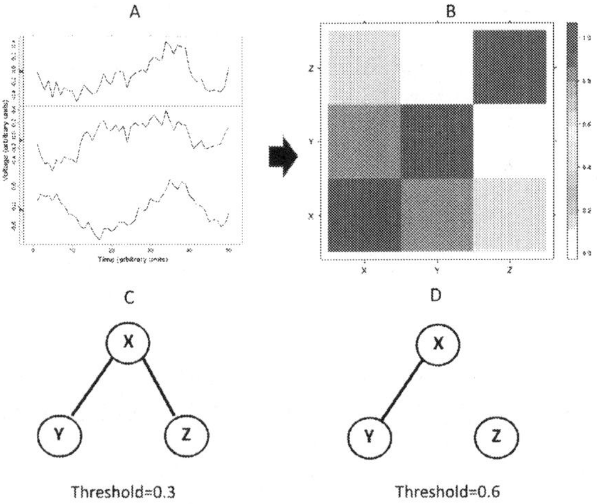

Figure 3. Network construction from simulated time series: (A) three correlated time series, X, Y and Z, (B) correlation (Pearson) matrix of time series of panel A, (C) network corresponding to the correlation matrix of panel B (threshold = 0.3) and (D) network corresponding to the correlation matrix of panel B (threshold = 0.6).

In **Figure 3**, the basic steps of constructing a simple network from simulated time series are displayed. Three synthetic time series X, Y and Z are shown in **Figure 3A**. With the objective to know whether any kind of functional connectivity exists between these time series, a synchronization measure (see Section 2.3) is calculated. In this way, an estimation of the potential relations between these recordings is assessed. **Figure 3B** shows a simple correlation estimate, Eq. (1), between these three time series. This figure shows the existence of a correlation close to 0.8, a rather intense value, between X and Y time series; the correlation between X and Z is close to 0.5, and between Y and Z is close to 0.2. Note that the correlation calculation accomplished in this way provides connectivity between any pair of time series even for the cases with very low values, giving rise to a fully connected network. This means that apparently uncorrelated time series, with a correlation values close to 0, will nonetheless be linked. Several methods can be used to eliminate these weak links. The first and more obvious is to select only those statistically significant correlations. On the contrary, one can choose an arbitrary threshold and eliminate those correlations below this particular value. This is what is displayed in **Figure 3C** where a threshold equal to 0.3 has been used. The basic network constructed in this way possesses only two links, X-Y and X-Z, because the link Y-Z, equal to 0.2, has been eliminated by the used threshold of 0.3. Whether a more stringent threshold is used, for instance equal to 0.6 as in **Figure 3D**, one of the former links is also discarded, X-Z, and the new network possess only one link, X-Y and one isolated node, Z. We have constructed in this way a non-connected network, because it possesses isolated nodes.

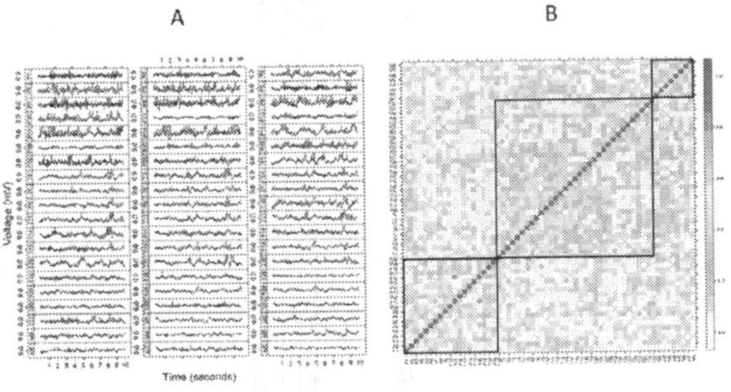

Figure 4. Network construction from neurophysiological time series: (A) actual time series from subdural electrodes and (B) matrix of absolute values of the correlation (Pearson) measure. Solid black squares delimit intra-area interactions.

A more complex and real example is depicted in **Figure 4**, taken from a typical neurophysiological recording of subdural electrodes. The set of electrodes comprises two subdural grids of 5×4 and 4×8 electrodes and a strip of 1×8 electrodes. A typical recoding lasting 10 s is displayed in **Figure 4A** and the correlation matrix in panel B. Note that all of the correlations values are positive. This is so because we have plotted, and also used, the absolute value of the correlation estimate. No matter in which "direction" the relation exists as long as it exists. It is easy to recognize in this figure the approximate boundaries of intra-area correlations (black solid lines). As mentioned above, the next step is in the selection of a threshold with the objective of simplifying the network. In **Figure 5A** and **B**, two examples of thresholded correlation matrices with 0.2 and 0.5, respectively, are displayed. Using this information the last step is to construct the network as the ones displayed in **Figure 6**. In the first case, **Figure 6A**, the correlation matrix with a threshold of 0.2 was used to construct the network. Due to the low value of threshold employed, 0.2, too many links populate the network. When greater threshold value is employed, as 0.5, only those stronger links remain (**Figure 6B**). As a final remark, note that both networks displayed in **Figure 6** seem to be of the kind unweighted (see Section 3.1), because no apparent differences exist between the links' width. However, every link has a weight associated with it, which is given by the corresponding correlation value. Thus, from this point on one can choose between two different scenarios to work, whether on an unweighted or weighted network. Different network properties and measures are explained in the next sections.

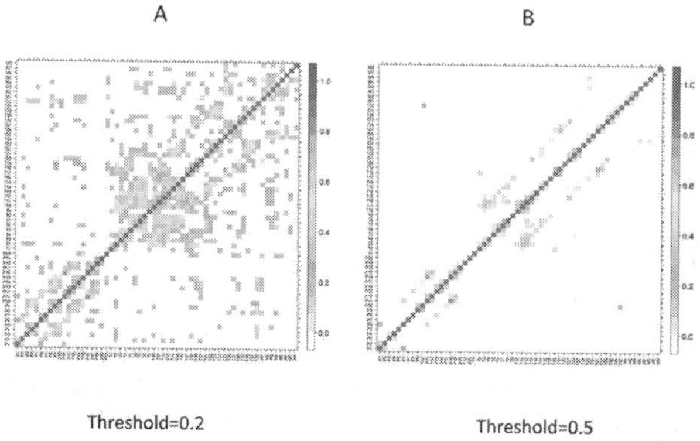

Figure 5. Network construction from neurophysiological time series: (A) Filtered (thresholded) correlation matrix when a threshold equal to 0.2 is applied and (B) filtered (thresholded) correlation matrix when a threshold equal to 0.5 is applied.

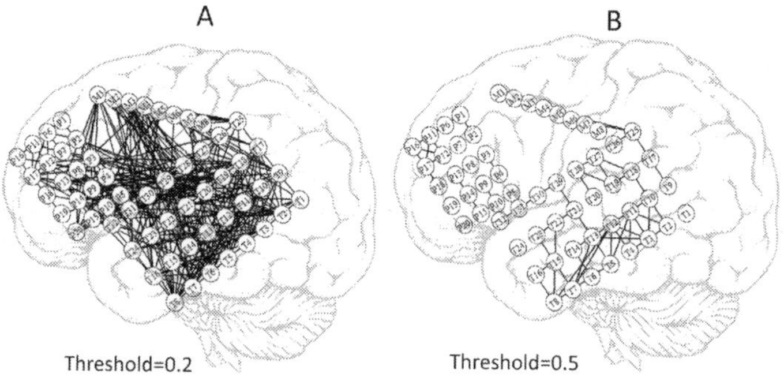

Figure 6. Network construction from neurophysiological time series: (A) network from time series of **Figure 4**, derived from the correlation matrix of **Figure 5A** and **(B)** network from time series of **Figure 4**, derived from the correlation matrix of **Figure 5B**.

3.3. Centrality measures

Centrality measures aim to study the nodes' characteristics and their relevance inside the network (**Figure 7**). Centrality measures characterize hubs nodes inside the network, a fact of surmount importance regarding seizure onset and spread. There are several centrality measures classified accordingly with the characteristics they measure, but we will only present here the most used in the literature, namely: degree, betweenness and local synchronization strength (for review see [10,26,67–69]). All these measures can be calculated for both weighted and unweighted graphs, although we only include here those corresponding to unweighted graphs. When appropriate, weighted definitions are going to be explained. For an extended review of weighted definitions, see [69].

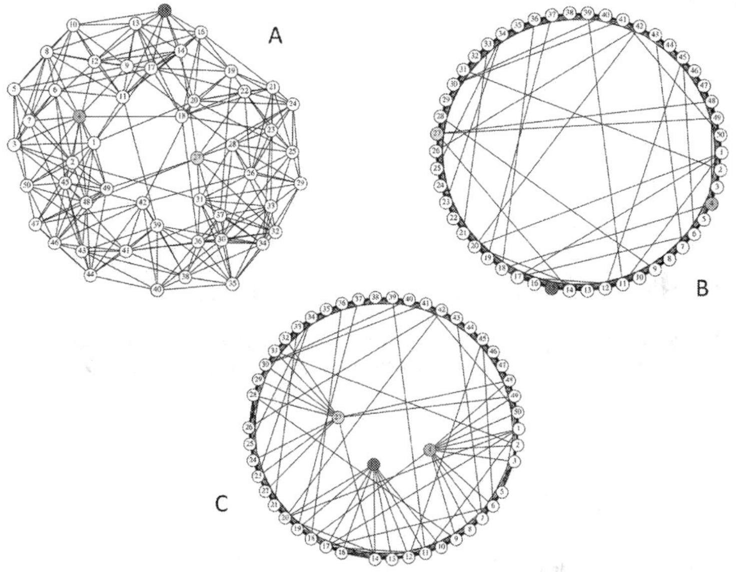

Figure 7. Centrality measures in a simulated network. In red, the node with maximum clustering coefficient. In blue, the node with maximum degree and, in yellow, the maximum betweenness. (a) Disordered network. (b) ordered network according to numeration, i.e. nodes are numbered according to their neighborhood. (c) ordered network according to numeration highlighting nodes with maximum values of centrality measures.

3.3.1. Concepts

As shown in Section 3.1, an unweighted network is fully described by the adjacency matrix a_{ij}. The number of links of a particular network node is called *node's degree*. The degree k_i of a node i is the number of edges that connect to other nodes

$$k_i = \sum_{j \in N} a_{ij}, \tag{3}$$

where N is number of network's nodes and a_{ij} are the binary element of the adjacency matrix. In a directed graph, the node's degree corresponds to the sum of ongoing k_i^{in} and outgoing

k_i^{out} edges, $k_i = k_i^{in} + k_i^{out}$. The importance of a node inside the network is directly linked to its degree level, i.e. a node with high degree possesses a great number of connections with other nodes of the network, increasing substantially connectivity with the rest of the network.

Betweenness can be defined as the capability of a node to facilitate the communication across the network. Betweenness of a node is defined as the number of shortest pathways that pass through this node. The betweenness b_i of a node i is

$$b_i = \sum_{j,k \in N, j \neq k} \frac{n_{jk}(i)}{n_{jk}} \tag{4}$$

being $n_{jk}(i)$ the number of shortest pathways passing through node i, and n_{jk} the total number of shortest pathways between nodes j and k. The betweenness concept is also applicable to edges, being the number of shortest pathways passing through a particular edge. A node with high betweenness has a great influence in the communication between other network nodes serving itself as an in-between relay. A node with a high degree may have a high betweenness; however, a node with high betweenness may have a low degree if it is located in a strategic position in the network.

Local synchronization, also known as *strength,* is the sum of weights of a particular node with its first neighbors divided by its degree. Thus, the local synchronization is represented as [70]

$$s_i = \frac{1}{n_i} \sum_{j=1}^{n_i} w_{ij}, \tag{5}$$

being w_{ij} the synchronization value between nodes i and j, and n_i is the number of first neighbors of node i. Local synchronization gives an idea of the contribution of each node to the total synchronization activity; a node with a higher local synchronization will contribute greatly to the global synchronization and perhaps to determine role in a network as a hub.

Clustering coefficient is defined as the proportion of neighbors' nodes that are also neighbor one of each other, characterizing the local connectedness in a network. According to [71], the clustering properties can be overestimated if weights are not considered when calculating clustering coefficient. So, the weighted clustering coefficient of node i is

$$c_i = \frac{1}{s_i(k_i - 1)} \sum_{j,k} \frac{w_{ij} + w_{ik}}{2} a_{ij} a_{ih} a_{jh}, \tag{6}$$

where k_i is the degree of i, s_i is its strength, a_{ij} are the binary elements of the adjacency matrix and w_{ij} are the weights between nodes i and j. The clustering is considered a segregation parameter because it describes the existence of specialized nodes, i.e. higher clustering coefficient means a higher connection density in the local subnetwork that surrounds the given node.

The following measure is not exactly a centrality measure, but nonetheless its importance has been explained in the following sections. The *Shortest path lengths* is the smallest number of edges that connect two nodes i and j, that is

$$d_{ij} = \sum_{a_{uv} \in g_{i \leftrightarrow j}} a_{uv}, \tag{7}$$

where $g_{i \leftrightarrow j}$ is the shortest path between i and j across the network nodes. The shortest path lengths is considered a measure of integration because it describes connectivity between distant nodes. This is why shortest path lengths is more informative as a global parameter, providing information about the global network integration through the average path length (see Section 3.5).

3.3.2. *Applications*

The most basic application of centrality nodes in epilepsy is in localizing areas involved in seizures. In [70] an approximation to this critical issue was carried out with the objective to assess whether nodes with high local synchronization participate in seizures in one way or another. In this study, functional connectivity was evaluated during intra-operatory ECoG by using three different measures: cross-correlation, phase synchronization and mutual information (Section 2.3), with a better performance of the two firsts [70]. Those cortical areas covered by electrodes with higher local synchronization seem to be deeply involved in seizures appearance because when these areas were resected during the surgery, patients remained without post-operative seizures. These results agree with other groups that observed higher synchrony in seizure onset zone [6,46,72].

Moreover, those areas with higher local synchronization also display low temporal variability, suggesting that their stability is also critical at the time to be involved in seizure generation [15]. It is argued [15] that the existence of particular areas with both high local synchronization and low temporal variability increase *seizurability*, i.e. the capability of the network to seize. On the contrary, no correlation could be established between these high synchronization areas and seizure onset zones. Altogether, these results seem to favor the hypothesis about seizures generation, which postulates that desynchronization is a preexistent state of the cortical areas and a transient synchronization help to spread the seizures.

The above-reported findings were all accomplished during the interictal period. However, long-run analysis carried out on subdural electrodes was also performed with the objective to explore the dynamics of these high local synchronization areas. In one study [18] of a patient with partial seizures – with and without secondary generalizations – a similar analysis was carried out. The analysis during partial seizures revealed temporal changes in those areas with higher local synchronization. Both types of seizures start in areas with high interictal local synchronization and both seizures present similar patterns in the first part of the seizure. These data support the hypothesis of the role of interictal local synchronization areas in seizure formation.

To summarize, altogether these data demonstrate how the centrality measures are a valuable tool for the analysis of invasive neurophysiological recordings, since it makes possible the characterization of the cortical dynamics in epilepsy patients. Specifically, these kinds of works prove the existence of stable local synchronization areas, which are involve in seizures generation because if they are surgically removed, the patients present better outcomes. The existence of these local synchronization areas implies the existence of areas with no high connectivity but also very intense and temporally stable local synchronization. It seems therefore that the cortex of epilepsy patients presents a highly heterogeneous connectivity, something that has been proved by immunohistochemistry, genetics and electrophysiological studies [73–75]. In addition, computational studies have shown that those areas of high local synchrony facilitate the global synchronization processes [24]. The stability of those local synchrony areas has been demonstrated that induce a change in network topology, to a small-world architecture (see Section 3.5.1), simplifying the synchronous activity between regions and favoring seizure onset.

3.4. Community structures

Detection of community structures is aimed to study the topological organization of a network accordingly with its subnetworks [26]. Community structures characterize clusters of tightly connected nodes inside the entire network. There are several community measures, the most used are motifs and modularity [69]. In the last years several methods to detect community structures have been published with great differences in both performances and capabilities, some of them using highly sophisticated algorithms [26].

3.4.1. Concepts

Modularity determines how well a given partition or division in a complex network corresponds to a natural or expected sub-division, i.e. which groups of nodes are more connected between them than with other nodes of the network. Thus modularity is defined as [69]

$$Q = \frac{1}{2m}\sum_{i,j}(a_{ij} - \frac{k_i k_j}{2m})\sigma(c_i, c_j), \tag{8}$$

where m is the number of edges, a_{ij} is the element of the adjacency matrix, k_i and k_j are the degree of node i and j, respectively; c_i is the type (or component) of i, c_j that of j, the sum goes over all i and j pairs of vertices, and σ (x, y) is 1 if $x = y$ and 0 otherwise.

In addition to the above-mentioned definition of community, an important issue in community structures is the method to be used to find them. Unlike other network parameters that are computed exactly, community structure calculations are obtained by optimizing algorithms.

Although several and highly sophisticated algorithms to calculate community exist [26], one of them, based on cutting a hierarchical tree, will be explained in the following section to understand its importance. A *Hierarchical clustering* algorithm is a method based on the construction of a hierarchy or tree based in similarities. The lower branches of the tree are composed of those more closely related nodes. In order to construct a hierarchical tree, a "distance" between objects is firstly defined and then, an ordering of distances is performed with the aim to construct the tree. A typical measure of distance is the so-called Gower distance

$$d_{(i,j)} = \sqrt{2(1-\rho_{ij})}, \tag{9}$$

based on the Pearson correlation (Eq. (1)) ρ_{ij}. In doing so, elements highly correlated with ρ_{ij} close to 1 attain distances close to 0. Then, one can construct a hierarchical tree by grouping those elements with similar distances. The process of assignations of nodes to a group or cluster can be done by combining nodes into groups by using an agglomerative method, or by separating groups into smaller ones by means of divisive methods. The single-linkage, complete-linkage and the average-link are the most common agglomerative methods. Once a hierarchical tree, also known as dendrogram, is constructed, the last step is simply to cut the tree at a particular level, obtaining the clusters or communities. A simple example can be observed in [26].

3.4.2. Applications

Communities or cluster detection has been successfully applied to neurophysiological data coming from scalp and FOE [28,49]. In these works, clustering detection was done by hierarchical clustering using a single-linkage clustering method. The aim of clustering detection was used to detect exactly the opposite, that is, those electrodes which do not belong to any cluster, or declustered nodes. Detection of declustered electrodes is important because it allows to correctly lateralize the ipsilateral side to seizures during the interictal period in TLE patients [28,49]. This calculation is performed through a lateralization index, quantifying which lobe possesses more declusterized/desynchronized electrodes. However, some controversy exists about the synchronization level of the ipsilateral side, since there are works describing higher synchronization in the ipsilateral side [76,77]. This controversy could be explained because of the different kinds of invasive electrodes, and different recording areas, used in those studies.

Community structures detection has been also used over subdural electrodes as in [18] in order to calculate the modularity. In this work it is shown that the value of modularity decreases during the seizures as compared with the preictal levels. Similar results are found by using a combination of scalp and FOE.

These results provide evidence that community detection is a promising diagnostic tool for network epilepsy, since it could help to determine the side of the seizure in TLE patients in semi-invasive interictal recordings of 1 or 2 h, saving time and using a less disturbing and a cheaper technique. Community detection can also help to determine the ictal dynamics through the analysis of modularity in ECoG and scalp and FOE. In addition to the above-mentioned community structures measures, others parameters can also help to analyze communities. Those are the network resilience measures, which define the resistance of a network to remove the random or critical nodes, as occurs in epilepsy surgery, for review see [69].

3.5. Topological organization

The epileptic network can also be characterized by changes in its topological organization which is reflected in the changes of other several parameters. One way to quantifying these changes is through averages, over the whole network, of certain measures, as for instance the shortest path length (Eq. (7)). Another way consists of transforming the entire network into a simplified network, called the minimum spanning tree, in order to study the main properties of the actual network by using the simplified one, instead.

3.5.1. Concepts

Average shortest path length (APL) is a network parameter employed to provide information of how fast or slow is the communication transfer through the network nodes, i.e. the average number of steps along the shortest paths through the network nodes. APL is defined as

$$APL = \frac{1}{n(n-1)} \sum_{i \neq j} d(v_i, v_j), \tag{10}$$

being n the number of nodes and $d(v_i, v_j)$ the shortest path lengths between nodes v_i and v_j, calculated by Eq. (7).

The *density of links* (DoL) is the ratio between the actual number of links and all possible links of the network. It provides information of how globally connected is the network. DoL is calculated as follows

$$DoL = \frac{\#of\ existing\ links}{\#of\ possible\ links}. \tag{11}$$

Both APL and DoL are related, i.e. a high DoL implies a low APL because a higher number of links entails higher possibilities to interconnect nodes and reaching this connection by using shortest paths. On the other hand, one cannot assume that a low APL implies a high DoL, because the presence of long-distance connected nodes favor the low APL without a high number of links, this is measured by the next parameter, the clustering coefficient.

Average clustering coefficient (ACC) measures how well neighbors nodes of a particular node are connected between them, characterizing local connectedness. ACC is defined as

$$ACC = \frac{1}{n}\sum_{i=1}^{n} c_i, \tag{12}$$

being n the number of nodes and c_j clustering coefficient of node i, calculated by Eq. (6).

The use of these three parameters allows a broad classification of networks in three groups: random, small-world and scale-free networks. *A random network* is a network with a uniform distribution of links, i.e. every node is connected with other every network's node with uniform probability. On the contrary, *small world network* is that network in which nodes have high local connectivity, i.e. high clustering coefficient, and some nodes also have long-distance connections. This last property represents that these small-world networks have a small average path length. Lastly, a *scale-free network* is that network in which most of the nodes have few local connections but some of them, called hubs, have a high number of connections. In the context of the average parameters, all three types can also be defined: random networks presented a low ACC and APL, regular networks showed a high ACC and low APL, small-world presented a high ACC and a low APL and scale-free presented similar characteristics to the small-world network.

The *minimum spanning tree (MST)* is a simplification of the full network into another, simpler, network. It is a tree without closed paths, i.e. from a node i to a node j always exists a unique path. The MST is obtained by the construction of a matrix based on the distance matrix, in the same way the construction of a hierarchical tree is done (see Section 3.4.1). The MST possesses the feature of retaining the more important links in the original network under the simplest topology.

3.5.2. Applications

To consider the topological aspects of networks, especially their changes, is of critical importance in its relation with epilepsy as it was demonstrated in several works [12,17,78]. A typical example of the application of network changes during seizures was recently presented by Vega-Zelaya et al. [18]. In this work differences in the network structure, constructed upon subdural recordings, between preictal, ictal and postictal stages were found. It is shown that after seizure onset of a partial seizure with secondary generalization, a decrease in both the modularity and the APL jointly with an increase in the DoL and the ACC exists.

Recently, the study of the transition from the preictal to the ictal period, recorded in scalp and FOE, reported that in 72% of the cases, an increase in the DoL during seizures with a decrease in the APL in 68% of cases exists. This fact suggests an increase in connectivity in the underlying functional network likely provoked by the small-world ictal architecture [58]. Although other works show a shift toward a regular network during the preictal-ictal transition [12,13,16], the difference could be explained by the use of FOE, which records electrical activity at the extrahippocampal areas [28]. In this regard, Mormann et al. [79] has revealed a different synchronization levels in the entorhinal cortex and hippocampus areas. Vega-Zelaya et al. [58] also reported an imbalance between ipsilateral and contralateral side, resulting in lower DoL and ACC and a higher APL in the ipsilateral side than the contralateral one. These results are also supported by fMRI studies [77,80], and even by results of studies of patients with extratemporal seizures, which presented lower connectivity in the seizure area [81].

The application of topological parameters to the analysis of epilepsy turned out to be a great tool for network characterization, either in ECoG, scalp or FOE recordings. It has uncovered the dynamics and connectivity of the brain in TLE as well in the extratemporal ones. In fact, these works showed a lack of connectivity in the seizure area. In particular, in TLE the lack of connectivity in the ipsilateral side could be somehow related to the ipsilateral impairment found in the community structures analysis of FOE [49]. Moreover, loss of connectivity within specific network structures has been involved in seizure generation [82], and computational models has revealed that connections deletion increases seizure likelihood [83].

Also, the use of the MST has provided to be of great importance in describing different types of functional architecture [68]. In that work, the MST was used to simplify the underlying network with the objective to localize particular areas of interest, as the node's degree, betweenness and local synchronization. As it was shown, critical nodes in the MST are those with highest local synchronization. Moreover, when different types of critical parameters, as maximum local synchronization, maximum betweenness, concur at a particular cortical area, resection of these area correlates with a goof post-operative outcome. **Figure 8** displays three different MST architectures with the critical nodes marked in each case. In the last case, **Figure 8C**, the three nodes with maximal local synchronization, betweenness and degree concur at the same location, that is node #2. During the surgery, a minimal cortical resection involved this cortical area and the patient remained free of seizures [68].

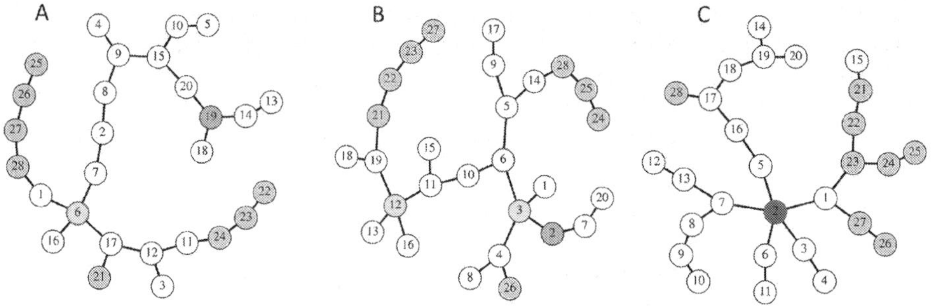

Figure 8. Centrality measures and minimum spanning tree (MST) constructed from ECoG data from two different locations. (A, B, C) Represent three examples from three patients. Electrodes are represented by gray circles (location 1) and white circles (location 2). The node with maximun local synchronization is represented by a cyan circle, the node with maximun degree by a yellow circle and The node with maximun betweenness by a red-border circle. Magenta circle represents superposition of the three centrality measures in the same node.

4. Conclusions

Although the advent of digital computation has revolutionized the whole realm of biomedical signal acquisition and preprocessing, particularly of scalp and invasive EEG, the clinical analysis and interpretation has not accompanied the rapid development experienced by the technological counterpart. Many of the current epilepsy-related protocols performed in the specialized centers around the world still rely on purely observational analyses of raw signals, depending subjectively on the neurophysiologist background and capabilities. Visual technique developed during the early years of epileptology, when computers were not yet available, has rooted so deeply in the clinical practice that it seems it could only be replaced by very efficient and objective techniques. In doing that, the new techniques should demonstrate that they performed at least as well as the classical ones. Moreover, they should provide a new framework under which the traditional view could be encompassed, but uncovering new concepts and models. Probably the little familiarity of neurophysiologists with numerical methods would help to explain to some extent the sparse use in clinical practice.

We feel that in the near future the complex network approach of epileptic processes will fit both requirements of being an encompassing framework under which the epileptogenesis and seizure spread could be fully understood and more important, it could become into a reliable therapeutic methodology to evaluate drug resistant epileptic patients, at least partially. As we have shown in this chapter, the traditional zone-oriented approach, with the seizure onset zone occupying the central position, is now shifting toward a network perspective, where network critical properties are becoming more relevant. This new framework would be applied to idiopathic or cryptogenic epilepsies, although symptomatic epilepsies (i.e. tumor induced, dysplasias, cavernomas or some types of post-traumatic seizure) probably remain better explained by the old or classical vision. In this regard, the graph theory jointly with the concept of functional connectivity is bringing new ways of thinking about the epilepsy.

Synchronization, of critical importance in constructing functional networks, has also evolved since its first use describing epileptic processes. Although epilepsy was initially related with hypersynchronization processes, today it is known that desynchronization also plays a dramatic role in the seizures dynamics. Moreover, as we have revised in this chapter, not only a single concept of synchronization exists but, on the contrary, several types of them are described.

One point which should be highlighted is the great difference existing between the traditional neurophysiological signal analysis in the field of epilepsy and the one performed under the network approach. As it is well known, interictal analysis is performed mainly in search of epileptogenic activity, i.e. sharp waves and spikes. On the contrary, synchronization of the full signals is made in construction networks.

Besides the novel knowledge provided by synchronization, the recent connectome concept has also added evidences to support the epilepsy as a truly network pathology. The analysis of EEG recordings from epilepsy patients reveals dynamical process in which network changes favor seizure creation. Centrality parameters show the presence of stable local synchronized areas that present a small-world architecture; community structures analysis revealed an increase in the ipsilateral side of desynchronized clusters, which can be related to the lower connectivity of the ipsilateral side as shown by topological parameters. All these results are at some points controversial mainly due to the different electrodes used for recordings and the small number of studies. However, they provide critical dynamical evidences to understand the underlying mechanism of epilepsy.

Although today we have the fragmented knowledge of the epilepsy as a complex network pathology, the new advances in network-related methodologies and the growing literature of its application to biomedical signals make network analyses a promising and yet powerful tool for epilepsy research with potential applications either in the diagnostic or in the therapeutic realms.

Acknowledgements

This work was funded by grants from Instituto de Salud Carlos III, through PI10/00160 and PI12/02839 and partially supported by F.E.D.E.R., from PIP CONICET 11220150100133CO and Mutua Madrileña. A.S.-G. is the recipient of a postdoctoral fellow from Mutua Madrileña.

Author details

Ancor Sanz-García[1], Rafael G. de Sola[2], Lorena Vega-Zelaya[1,3], Jesús Pastor[1,3] and Guillermo J. Ortega[1,2*]

*Address all correspondence to: gjortega.hlpr@salud.madrid.org

1 Unidad de Neurocirugía de la Epilepsia, Instituto de Investigaciones Biomédicas Hospital de la Princesa, Madrid, Spain

2 Neurocirugía, Hospital Universitario de la Princesa, Madrid, Spain

3 Neurofisiología Clínica, Hospital Universitario de la Princesa, Madrid, Spain

References

[1] Villanueva V, Sánchez-álvarez JC, Peña P, Puig JS, Caballero-Martínez F, Gil-Nagel A. Treatment initiation in epilepsy: an expert consensus in Spain. Epilepsy Behav. 2010;19:332–42. DOI:10.1016/j.yebeh.2010.07.016

[2] Thom M, Gary WM, Cross JH, Bertram EH. Mesial temporal lobe epilepsy: how do we improve surgical outcome? Ann Neurol. 2010;68:424–34. DOI: 10.1002/ana.22142

[3] Bertram EH, Xing-Zhang D, Mangan P, Fountain N, Rempe D. Functional anatomy of limbic epilepsy: a proposal for central synchronization of a diffusely hyperexcitable network. Epilepsy Res. 1998;32:194–205. DOI:10.1016/S0920-1211(98)00051-5

[4] Bartolomei F, Wendling F, Bellanger J, Regis J, Chauvel P. Neural networks involved in temporal lobe seizures: a nonlinear regression analysis of SEEG signals interdependencies. Clin Neurophysiol. 2001;112:1746–60. DOI:10.1016/S1388-2457(01)00591-0.

[5] Spencer SS. Neural networks in human epilepsy: evidence of and implications for treatment. Epilepsia. 2002;43:219–27. DOI:10.1046/j.1528-1157.2002.26901.x

[6] Bartolomei F, Wendling F, Regis J, Gavaret M, Guye M, Chauvel P. Pre-ictal synchronicity in limbic networks of mesial temporal lobe epilepsy. Epilepsy Res. 2004;61:89–104. DOI: 10.1016/j.eplepsyres.2004.06.006

[7] Vega-Zelaya L, Torres CV, Garnes-Camarena O, Ortega GJ, García-Navarrete E, Navas M, Sola RG, Pastor J. Electrocorticographic evidence and surgical implications of different physiopathologic subtypes of temporal epilepsy. Clin. Neurophysiol. 2014;125:2349–57. DOI: 10.1016/j.clinph.2014.03.027

[8] Berg AT, Scheffer IE. New concepts in classification of the epilepsies: entering the 21st century. Epilepsia. 2011;52:1058–62. DOI: 10.1111/j.1528-1167.2011.03101.x

[9] Bernhardt BC, Bonilha L, Gross DW. Network analysis for a network disorder: the emerging role of graph theory in the study of epilepsy. Epilepsy Behav. 2015;50:162–70. DOI:10.1016/j.yebeh.2015.06.005

[10] Sporns O, Tononi G. Structural determinants of functional brain dynamics. In: Jirsa VK, McIntosh AR, editors. Handbook of brain connectivity. Berlin Heidelberg: Springer; 2007. p. 117–147. DOI: 10.1007/978-3-540-71512-2

[11] Rosenow, F, Luders, H. Presurgical evaluation of epilepsy. Brain. 2001;124:1683–700. DOI: 10.1093/brain/124.9.1683

[12] Ponten SC, Bartolomei F, Stam CJ. Small-world networks and epilepsy: graph theoretical analysis of intracerebrally recorded mesial temporal lobe seizures. Clin Neurophysiol. 2007;118:918–27. DOI: 10.1016/j.clinph.2006.12.002

[13] Schindler KA, Bialonski S, Horstmann MT, Elger CE, Lehnertz K. Evolving functional network properties and synchronizability during human epileptic seizures. Chaos. 2008;18:033119. DOI: 10.1063/1.2966112

[14] Kramer MA, Eden UT, Kolaczyk ED, Zepeda R, Eskandar EN, Cash SS. Coalescence and fragmentation of cortical networks during focal seizures. J Neurosci. 2010;30:10076–85. DOI: 10.1523/JNEUROSCI.6309-09.2010

[15] Palmigiano A, Pastor J, de Sola RG, Ortega GJ. Stability of synchronization clusters and seizurability in temporal lobe epilepsy. PLoS One. 2012;7:e41799. DOI: 10.1371/journal.pone.0041799

[16] Diessen E, Diederen SJ, Braun KP, Jensen FE, Stam CJ. Functional and structural brain networks in epilepsy: what have we learned? Epilepsia. 2013;54:1855–65. DOI: 10.1111/epi.12350

[17] van Diessen E, Zweiphenning WJ, Jansen FE, Stam CJ, Braun KP, Otte WM. Brain network organization in focal epilepsy: a systematic review and meta-analysis. PLoS One. 2014;9:e114606. DOI:10.1371/journal.pone.0114606

[18] Vega-Zelaya L, Pastor JE, de Sola RG, Ortega GJ. Inhomogeneous cortical synchronization and partial epileptic seizures. Front Neurol. 2014;24:187. DOI: 10.3389/fneur.2014.00187

[19] Centeno M, Carmichael DW. Network connectivity in epilepsy: resting state fMRI and EEG-fMRI contributions. Front Neurol. 2014;5:93. DOI: 10.3389/fneur.2014.00093

[20] de Lanerolle NC, Lee TS, Spencer DD. Histopathology of human epilepsy. In: Noebels JL, Avoli M, Rogawski MA, et al., editors. Jasper's basic mechanisms of the epilepsies [Internet]. 4th edition. Bethesda (MD): National Center for Biotechnology Information (US); 2012.

[21] Pikovsky A, Rosenblum M, Kurths J. Synchronization: a universal concept in nonlinear sciences. Cambridge Nonlinear Science Series. New York: Cambridge University Press; 2001.

[22] Press WH, Teukolsky SA, Vetterling WT, Flannery BP. Numerical recipes, The art of scientific computing, 3rd edition, New York: Cambridge University Press; 2007.

[23] Stam CJ. Nonlinear dynamical analysis of EEG and MEG: review of an emerging field. Clin Neurophysiol. 2005;116:2266–301. DOI:10.1016/j.clinph.2005.06.011

[24] Pastor J, de Sola RG, Ortega GJ. Hyper-synchronization, de-synchronization, synchronization and seizures. In: Stevanovic D, editor. Epilepsy – histological, electroencephalographic and psychological aspects. Rijeka, Croatia: InTech; 2012. DOI: 10.5772/31004

[25] Cover TM, Thomas JA. Elements of information theory, 2nd edition. Chichester: Wiley; 2011.

[26] Boccaletti S, Latora V, Moreno Y, Chavez M, Hwang D-H. Complex networks: structure and dynamics. Phys Rep. 2006;424:175–308. DOI: 10.1016/j.physrep.2005.10.009

[27] Penfield W, Jasper H. Hypersynchrony. Epilepsy and the functional anatomy of the human brain. Boston: Little-Brown; 1954.

[28] Pastor J, Navarrete EG, Sola RG, Ortega GJ. Extrahippocampal desynchronization in nonlesional temporal lobe epilepsy. Epilepsy Res Treat. 2012;2012:823683. DOI: 10.1155/2012/823683

[29] Jiruska P, de Curtis M, Jefferys JG, Schevon CA, Schiff SJ, Schindler K. Synchronization and desynchronization in epilepsy: controversies and hypotheses. J Physiol. 2013;591:787–97. DOI: 10.1113/jphysiol.2012.239590

[30] Dichter MA, Ayala GF. Cellular mechanisms of epilepsy: a status report. Science. 1987;237:157–64. DOI: 10.1126/science.3037700

[31] Kandel ER, Schwartz JH, Jessell TM. Principles of neural science, 4th edition. New York: McGraw-Hill; 2000.

[32] McCormick DA, Contreras D. On the cellular and network bases of epileptic seizures. Annu Rev Physiol. 2001;63:815–46. DOI: 10.1146/annurev.physiol.63.1.815

[33] Timofeev I, Steriade M. Neocortical seizures: initiation, development and cessation. Neuroscience. 2004;123:299–336. DOI: 10.1016/j.neuroscience.2003.08.051

[34] Speckmann E-J, Elger CE, Gorji, A. Neurophysiologic basis of EEG and DC potentials. In: Schomer DL, Lopes da Silva, editors. Niedermeyer's electroencephalography: basic principles, clinical applications, and related fields, 6th ed. Philadelphia: Lippincott Williams & Wilkins; 2011.

[35] Noachtar S, Rémi J. The role of EEG in epilepsy: a critical review. Epilepsy Behav. 2009;15:22–33. DOI: 10.1016/j.yebeh.2009.02.035

[36] Walczak TS, Jayakar P, Mizrahi EM. Interictal Electroencephalography. In: Engel J, Pedley TA, editors. Epilepsy: a comprehensive textbook, 2nd edition. Philadelphia: Lippincott Williams & Wilkins; 2008.

[37] Traub RD, Bibbig A, LeBeau FEN, Buhl EH, Whittington MA. Cellular mechanisms of neuronal population oscillations in the hippocampus in vitro. Annu Rev Neurosci. 2004;27:247–78. DOI: 10.1146/annurev.neuro.27.070203.144303

[38] Morgan RJ, Soltesz I. Nonrandom connectivity of the epileptic dentate gyrus predicts a major role for neuronal hubs in seizures. Proc Natl Acad Sci U S A. 2008;16:6179–84. DOI: 10.1073/pnas.0801372105

[39] Zhang M, Ladas TP, Qiu C, Shivacharan RS, Gonzalez-Reyes LE, Durand DM. Propagation of epileptiform activity can be independent of synaptic transmission, gap junctions, or diffusion and is consistent with electrical field transmission. J Neurosci 2014;34:1409–19. DOI: 10.1523/JNEUROSCI.3877-13.2014

[40] Bourien J, Bartolomei F, Bellanger JJ, Gavaret M, Chauvel P, Wendling F. A method to identify reproducible subsets of co-activated structures during interictal spikes. Application to intracerebral EEG in temporal lobe epilepsy. Clin Neurophysiol. 2005;116:443–55. DOI: 10.1016/j.clinph.2004.08.010

[41] Engel J. Seizures and epilepsy. Philadelphia, PA: FA Davis; 1989.

[42] Pastor J, Menéndez de la Prida L, Hernando V, Sola RG. Voltage sources in mesial temporal lobe epilepsy recorded with foramen ovale electrodes. Clin Neurophysiol 2006;117:2604–14. DOI: 10.1016/j.clinph.2006.07.311

[43] Townsend TN, Ebersole JS. Source localization of electroencephalography spikes, In: Luders HO, editor. Textbook of epilepsy surgery. Boca Raton, Florida: CRC Press: 2008.

[44] Spencer SS, Spencer DD. Entorhinal-hippocampal interactions in medial temporal lobe epilepsy. Epilepsia. 1994;4:721–27. DOI: 10.1111/j.1528-1157.1994.tb02502.x

[45] Pastor J, Sola RG, Ortega GJ. Influence of paroxysmal activity on background synchronization in epileptic recordings. J Neurosci Methods. 2014;223:69–73. DOI: 10.1016/j.jneumeth.2013.11.027

[46] Mormann F, Lehnertz K, David P, Elger CE. Mean phase coherence as a measure for phase synchronization and its application to the EEG of epilepsy patients. Physica D. 2000;144:358–69. DOI: 10.1016/S0167-2789(00)00087-7

[47] Andrzejak R, Mormann F, Widman G, Kreuz T, Elger C, Lehnertz K. Improved spatial characterization of the epileptic brain by focusing on nonlinearity. Epilepsy Res. 2006;69:30–44. DOI: 10.1016/j.eplepsyres.2005.12.004

[48] Bettus G, Wendling F, Guye M, Valton L, Regis J, Chauvel P, et al. Enhanced EEG functional connectivity in mesial temporal lobe epilepsy. Epilepsy Res. 2008;81:58–68. DOI: 10.1016/j.eplepsyres.2008.04.020

[49] Ortega GJ, Peco IH, Sola RG, Pastor J. Impaired mesial synchronization in temporal lobe epilepsy. Clin Neurophysiol. 2011;122:1106–16. DOI: 10.1016/j.clinph.2010.11.001

[50] Schindler K, Leung H, Elger CE, Lehnertz K. Assessing seizure dynamics by analyzing the correlation structure of multichannel intracranial EEG. Brain. 2007;130:65–77. DOI: 10.1093/brain/awl304

[51] Babb TL, Wilson CL, Isokawa-akesson M. Firing patterns of human limbic neurons during stereoencephalography (SEEG) and clinical temporal–lobe seizures. Electroencephal Clin Neurophysiol. 1987;66:467–82. DOI: 10.1016/0013-4694(87)90093-9

[52] Netoff TI, Schiff SJ. Decreased neuronal synchronization during experimental seizures. J Neurosci. 2002;22:7297–307.

[53] Mormann F, Kreuz T, Andrzejak RG, David P, Lehnertz K, Elger CE. Epileptic seizures are preceded by a decrease in synchronization. Epilepsy Res. 2003;53:173–85. DOI: 10.1016/S0920-1211(03)00002-0

[54] Gastaut H, Roger J, Ouahchi S, Timsit M, Broughton R. An electro-clinical study of generalized epileptic seizures of tonic expression. Epilepsia. 1963;4:15–44. DOI: 10.1111/j.1528-1157.1963.tb05206.x

[55] Le Van Quyen M, Navarro V, Martinerie J, Baulac M, Varela FJ. Toward a neurodynamical understanding of ictogenesis. Epilepsia. 2003;44:30–43. DOI: 10.1111/j.0013-9580.2003.12007.x

[56] Wendling F, Bartolomei J, Bellanger J, Bourien J, Chauvel P. Epileptic fast intracerebral EEG activity: evidence for spatial decorrelation at seizure onset. Brain. 2003;126:1449–59. DOI: 10.1093/brain/awg144

[57] Milton JG, Chkhenkeli SA, Towle VL. Brain connectivity and the spread of epileptic seizures. In: Jirsa VK, McIntosh AR, editors. Handbook of brain connectivity. Berlin Heidelberg: Springer; 2007. p. 117–47. DOI: 10.1007/978-3-540-71512-2

[58] Vega-Zelaya L, Pastor J, de Sola RG, Ortega GJ. Disrupted ipsilateral network connectivity in temporal lobe epilepsy. PLoS One. 2015;10(10), e0140859.

[59] Bragin A, Wilson CL, Engel J. Chronic epileptogenesis requires development of a network of pathologically interconnected neuron clusters: a hypothesis. Epilepsia. 2000;41:S144–52. DOI: 10.1111/j.1528-1157.2000.tb01573.x

[60] Jiruska P, Csicsvari J, Powell AD, Fox JE, Chang WC, Vreugdenhil M, et al. High-frequency network activity, global increase in neuronal activity, and synchrony expansion precede epileptic seizures in vitro. J Neurosci. 2010;30:5690–701. DOI: 10.1523/JNEUROSCI.0535-10.2010

[61] Schevon CA, Ng SK, Cappell J, Goodman RR, McKhann G Jr, Waziri A, et al. Microphysiology of epileptiform activity in human neocortex. J Clin Neurophysiol. 2008;25:321–30. DOI 10.1097/WNP.0b013e31818e8010

[62] Schevon CA, Goodman RR, McKhann Jr G, Emerson RG. Propagation of epileptiform activity on a submillimeter scale. J Clin Neurophysiol. 2010;27:406–11. DOI: 10.1097/WNP.0b013e3181fdf8a1

[63] Stead M, Bower M, Brinkmann BH, Lee K, Marsh WR, Meyer FB, et al. Microseizures and the spatiotemporal scales of human partial epilepsy. Brain. 2010;133:2789–97. DOI: 10.1093/brain/awq190

[64] Le Van Quyen M, Martinerie J, Navarro V, Baulac And M, Varela FJ. Characterizing neurodynamic changes before seizures. J Clin Neurophysiol. 2001;18:191–208.

[65] van Drongelen W, Lee HG, Hereld M, Chen ZY, Elsen FP, Stevens RL. Emergent epileptiform activity in neural networks with weak excitatory synapses. IEEE Trans Neural Sys Rehab. 2005;13:236–41. DOI: 10.1109/TNSRE.2005.847387

[66] Sporns O, Tononi G, Kotter R. The human connectome: a structural description of the human brain. PLoS Comput Biol. 2005;1:e42. DOI: 10.1371/journal.pcbi.0010042

[67] Sporns O, Chialvo DR, Kaiser M, Hilgetag CC. Organization, development and function of complex brain networks. Trends Cogn Sci. 2004;8:418–25. DOI: 10.1016/j.tics.2004.07.008

[68] Ortega GJ, Sola RG, Pastor J. Complex network analysis of human ECoG data. Neurosci Lett. 2008;447:129–33. DOI: 10.1016/j.neulet.2008.09.080

[69] Rubinov M, Sporns O. Complex network measures of brain connectivity: uses and interpretations. Neuroimage. 2010;52:1059–69. DOI: 10.1016/j.neuroimage.2009.10.003

[70] Ortega GJ, Menendez de la Prida L, Sola RG, Pastor J. Synchronization clusters of interictal activity in the lateral temporal cortex of epileptic patients: intraoperative electrocorticographic analysis. Epilepsia. 2008;49:269–80. DOI: 10.1111/j.1528-1167.2007.01266.x

[71] Barrat A, Barthelemy M, Pastor-Satorras R, Vespignani A. The architecture of complex weighted networks. Proc Natl Acad Sci U S A. 2004;101:3747–52. DOI: 10.1073/pnas.0400087101

[72] Arnhold J, Lehnertz K, Grassberger P, Elger CE. A robust method for detecting interdependences: application to intracranially recorded EEG. Physica D. 1999;134:419–30. DOI: 10.1016/S0167-2789(99)00140-2

[73] Ferrer I, Oliver B, Russi A, Casas R, Rivera R. Parvalbumin and calbindin-D28k immunocytochemistry in human neocortical epileptic foci. J Neurol Sci. 1994;123:18–25. DOI: 10.1016/0022-510X(94)90198-8

[74] Marco P, Sola RG, Pulido P, Alijarde MT, Sanchez A, Ramon y Cajal S, et al. Inhibitory neurons in the human epileptogenic temporal neocortex. An immunocytochemical study. Brain. 1996;119:1327–47. DOI: 10.1093/brain/119.4.1327

[75] Menendez de la Prida L, Benavides-Piccione R, Sola RG, Pozo MA. Electrophysiological properties of interneurons from intraoperative spiking areas of epileptic human temporal neocortex. Neuroreport. 2002;13:1421–5.

[76] Bartolomei F, Chauvel P, Wendling F. Epileptogenicity of brain structures in human temporal lobe epilepsy: a quantified study from intracerebral EEG. Brain. 2008;131:1818–30. DOI: 10.1093/brain/awn111

[77] Bettus G, Guedj E, Joyeux F, Confort-Gouny S, Soulier E, Laguitton V, et al. Decreased basal fMRI functional connectivity in epileptogenic networks and contralateral compensatory mechanisms. Hum Brain Mapp. 2009;30:1580–91. DOI: 10.1002/hbm.20625.

[78] Wilke C, Worrell G, He B. Graph analysis of epileptogenic networks in human partial epilepsy. Epilepsia. 2011;52:84–93. DOI: 10.1111/j.1528-1167.2010.02785.x

[79] Mormann F, Osterhage H, Andrzejak RG, Weber B, Fernández G, Fell J, et al. Independent delta/theta rhythms in the human hippocampus and entorhinal cortex. Front Hum Neurosci. 2008;2:1–6. DOI: 10.3389/neuro.09.003.2008

[80] Pereira FR, Alessio A, Sercheli MS, Pedro T, Bilevicius E, Rondina JM, et al. Asymmetrical hippocampal connectivity in mesial temporal lobe epilepsy: evidence from resting state fMRI. BMC Neurosci. 2010;11:66. DOI: 10.1186/1471-2202-11-66.

[81] Burns SP, Santaniello S, Yaffe RB, Jouny CC, Crone NE, Bergey GK, et al. Network dynamics of the brain and influence of the epileptic seizure onset zone. Proc Natl Acad Sci U S A. 2014;111:E5321–30. DOI: 10.1073/pnas.1401752111

[82] Berg AT, Berkovic SF, Brodie MJ, Buchhalter J, Cross JH, Van Emde Boas W, et al. Revised terminology and concepts for organization of seizures and epilepsies: report of the ILAE Commission on Classification and Terminology, 2005–2009. Epilepsia. 2010;51:676–85. DOI: 10.1111/j.1528-1167.2010.02522.x

[83] Terry JR, Benjamin O, Richardson MP. Seizure generation: the role of nodes and networks. Epilepsia. 2012;53:e166–9. DOI: 10.1111/j.1528-1167.2012.03560.x

Chapter 5

Enhancing Estimates of Breakpoints in Genome Copy Number Alteration using Confidence Masks

Jorge Muñoz-Minjares, Yuriy Shmaliy and
Oscar Ibarra-Manzano

Additional information is available at the end of the chapter

http://dx.doi.org/10.5772/63913

Abstract

Chromosomal structural changes in human body known as copy number alteration (CNA) are often associated with diseases, such as various forms of cancer. Therefore, accurate estimation of breakpoints of the CNAs is important to understand the genetic basis of many diseases. The high-resolution comparative genomic hybridization (HR-CGH) and single-nucleotide polymorphism (SNP) technologies enable cost-efficient and high-throughput CNA detection. However, probing provided using these profiles gives data highly contaminated by intensive Gaussian noise having white properties. We observe the probabilistic properties of CNA in HR-CGH and SNP measurements and show that jitter in the breakpoints can statistically be described with either the discrete skew Laplace distribution when the segmental signal-to-noise ratio (SNR) exceeds unity or modified Bessel function-based approximation when SNR is <1. Based upon these approaches, the confidence masks can be developed and used to enhance the estimates of the CNAs for the given confidence probability by removing some unlikely existing breakpoints.

Keywords: copy number alterations, HR-CGH, SNP, breakpoints, confidence masks

1. Introduction

It is well known that the deoxyribonucleic acid (DNA) of a genome essential for human life often demonstrates structural changes [1–3] called genome copy number alterations (CNAs) [4–6], which are associated with disease such as cancer [7]. Analysis of the breakpoint locations in the CAN structure is still an important issue because it helps detecting structural alterations, load of alterations in the tumor genome, and absolute segment copy numbers. Thus, efficient estimators are required to extract information about the breakpoints with accuracy acceptable for medical needs. To produce CNA profile, several technologies have been developed such as comparative genomic hybridization (CGH) [8], high-resolution CGH (HR-CGH) [9], whole genome sequencing [10], and most recently single-nucleotide polymorphism (SNP) [11]. The HR-CGH technology is still used widely in spite of its low resolution [12]. It has been reported in [13] that the HR-CGH arrays are accurate to detect structural variations (SVs) at the resolution of 200 bp (*base pairs*). Most recently, the single-nucleotide polymorphism technology was developed in the study of Wang et al. [11] to provide high-resolution measurements of the CNAs. In spite of their high resolution, the modern methods still demonstrate the inability in obtaining good estimates of the breakpoint locations because of the following factors: (1) the nature of biological material (tumor is contaminated by normal tissue, relative values, and unknown baseline for copy number estimation), (2) technological biases (quality of material and hybridization/sequencing), and (3) intensive random noise. The HR-CGH and SNP profiles have demonstrated deficiency in detecting the CNAs, but noise in the detected changes still remains at a high level [14] and accurate estimators are required to extract information about structural changes.

In the HR-CGH microarray technique, the CNAs are often normalized and plotted as $\log_2 R/G = \log_2$ ratio, where R and G are the fluorescent Red and Green intensities, respectively [12]. The CNA measurements using SNP technologies are represented by the Log-R ratios (LRRs), which are the log-transformed ratios of experimental and normal reference SNP intensities centered at zero for each sample [14]. From the standpoint of signal processing, the following properties of the CNA function are of importance [15]:

- It is piecewise constant (PWC) and sparse with a small number of alterations on a long base-pair length.
- Constant values are integer, although this property is not survived in the log-R ratio.
- The measurement noise in the log-R ratio is highly intensive and can be modeled as additive white Gaussian.

The CNA estimation problem is thus to predict the breakpoint locations and the segmental levels with a maximum possible accuracy and precision acceptable for medical applications. In this work, we developed our methods to two types of cancer: B-cell chronic lymphocytic leukemia (B-CLL) and BLC primary breast carcinoma. Nevertheless, the methods were designed to any samples of cancer with the characteristics described above.

2. Methods

2.1. CNA model and problem statement

Consider a chromosome section observed with some resolution Δ, bp at M discrete breakpoints, $n \in [1, M]$. An example of the CNA probes with a single breakpoint and two segments is shown in **Figure 1**. Suppose that the copy numbers change at K breakpoints,

http://dx.doi.org/10.5772/63913

$1<i_1< \ldots <i_K<M$, united in a vector $\boldsymbol{I} = [i_1 i_2 \ldots i_K]^{\mathrm{T}}$. The measurement can thus be represented with a vector $\boldsymbol{y} \in \mathcal{R}^M$ as

$$y = \left[y_1 y_2 \ldots y_{i_1} y_{i_1} + 1 \ldots y_{i_2} \ldots y_{i_K} \ldots y_M \right]^T. \tag{1}$$

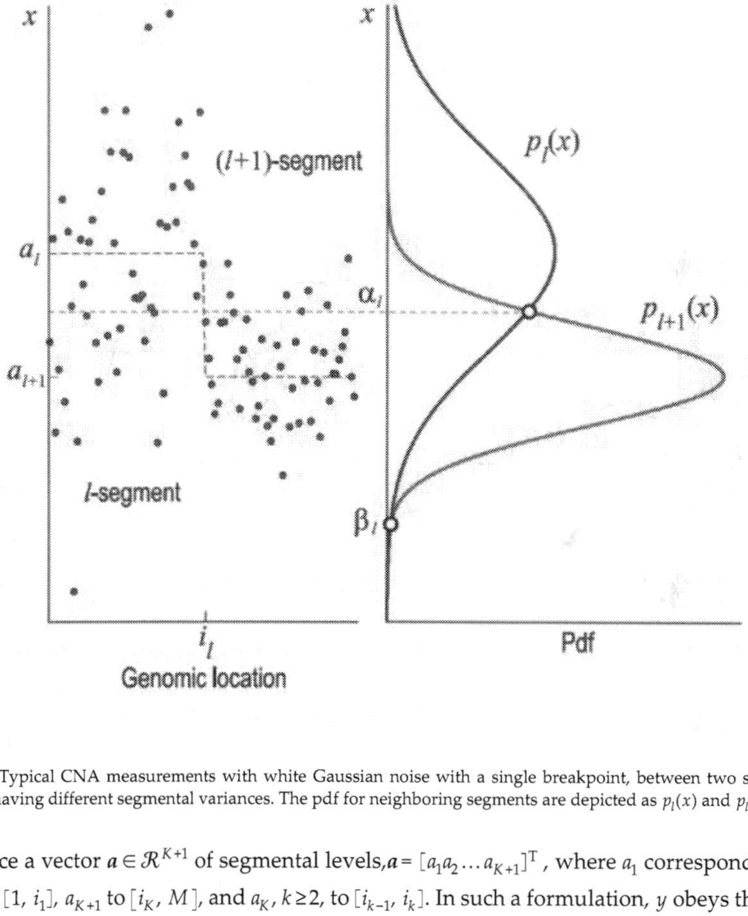

Figure 1. Typical CNA measurements with white Gaussian noise with a single breakpoint, between two segments l and $l+1$ having different segmental variances. The pdf for neighboring segments are depicted as $p_l(x)$ and $p_{l+1}(x)$.

Introduce a vector $\boldsymbol{a} \in \mathcal{R}^{K+1}$ of segmental levels, $\boldsymbol{a} = [a_1 a_2 \ldots a_{K+1}]^{\mathrm{T}}$, where a_1 corresponds to the interval $[1, i_1]$, a_{K+1} to $[i_K, M]$, and a_K, $k \geq 2$, to $[i_{k-1}, i_k]$. In such a formulation, y obeys the linear regression model

$$\boldsymbol{y} = \boldsymbol{A}(\boldsymbol{I})\boldsymbol{a} + \mathbf{v} \tag{2}$$

where the regression matrix $A \in \mathcal{R}^{M\times(K+1)}$ is sparse,

$$A = \left[A_1^T A_2^T \ldots A_{K+1}^T \right]^T, \tag{3}$$

having a component

$$A_k = \begin{bmatrix} 0 & \cdots & 1 & \cdots & 0 \\ 0 & \cdots & 1 & \cdots & 0 \\ \vdots & \ddots & \vdots & \ddots & \vdots \\ 0 & \cdots & 1 & \cdots & 0 \end{bmatrix}, \tag{4}$$

in which the kth column is filled with unity and all others are zeros. The number of the columns in A_k is exactly $K+1$. However, the number or the row depends on the interval $i_k - i_{k-1}$. Thus, the row-variant matrix (4) is $A_k \in \mathcal{R}^{(i_k - i_{k-1})\times(K+1)}$. Additive noise v in Eq. (2) is zero mean, $E\{\mathbf{v}\}=0$, and white Gaussian with the covariance $\boldsymbol{R}=\sigma_v^2\boldsymbol{I}$, where $\boldsymbol{I} \in \mathcal{R}^{M\times M}$ is an identity matrix and σ_v^2 is a know variance.

The CNA estimation problem is thus to predict the breakpoint locations and evaluate the segmental changes $\boldsymbol{x}=\boldsymbol{A}(I)\boldsymbol{a}$ with a maximum possible accuracy and precision acceptable for medical applications. The problem is complicated by short number of the probes in each neighboring segment and indistinct edges. Therefore, an analysis of the estimation errors caused by the segmental noise and jitter in the breakpoints is required.

2.2. Jitter probability in the breakpoints

Consider a typical genomic measurement of two neighboring CNA segments in white Gaussian noise with different segmental variances as shown in **Figure 1**. A constant signal changes from level a_l to level a_{l+1} around the *breakpoint* i_l. In the presence of noise, the location of i_l is not clear owing to commonly large segmental variances σ_l^2 and σ_{l+1}^2. As an example, the Gaussian noise probability density functions (pdfs) $p_l(x)$ and $p_{l+1}(x)$ are shown in **Figure 1** for $\sigma_l^2 > \sigma_{l+1}^2$. Let us notice that $p_l(x)$ and $p_{l+1}(x)$ cross each other in two points, α_l and β_l, provided that $\sigma_l^2 \neq \sigma_{l+1}^2$.

Now considerer N probes in each segment neighboring to i_l with an average resolution. We thus may assign an event A_{lj} meaning that measurement at point $i_l - N \leq j < i_l$ belongs to the lth segment. Another event B_{lj} means that measurement at $i_l \leq j < i_l + N - 1$ belongs to the $(l+1)$th segment. In our approach, we think that a measured value belongs to one segment if the probability is larger than if it belongs to another segment. For example, any measurement point in the interval between α_l and β_l (**Figure 1**) is supposed to belong to the $(l+1)$th segment.

http://dx.doi.org/10.5772/63913

Following **Figure 1** and assuming different noise variances σ_l^2 and σ_{l+1}^2, the events A_{lj} and B_{lj} can be specified as follows [16]:

$$\underset{i_l - N_l \le j < i_l}{A_{lj}} \text{ is} \begin{cases} (\alpha_l < x_j) \vee (x_j < \beta_l), & \sigma_l^2 > \sigma_{l+1}^2, \\ x_j > \alpha_l, & \sigma_l^2 = \sigma_{l+1}^2, \\ \alpha_l < x_j < \beta_l, & \sigma_l^2 < \sigma_{l+1}^2, \end{cases} \tag{5}$$

$$\underset{i_l \le j < i_l + N_l - 1}{B_{lj}} \text{ is} \begin{cases} \beta_l < x_j < \alpha_l & \sigma_l^2 < \sigma_{l+1}^2, \\ x_j < \alpha_l, & \sigma_l^2 = \sigma_{l+1}^2, \\ (x_j < \alpha_l) \vee (x_j > \beta_l), & \sigma_l^2 > \sigma_{l+1}^2. \end{cases} \tag{6}$$

Because each point can belong only to one segment, the inverse events are $\bar{A} = 1 - A_{lj}$ and $\bar{B} = 1 - B_{lj}$.

Events A_{lj} and B_{lj} can be united into two blocks: $\boldsymbol{A}_l = \{A_{l(i_l-N)} A_{l(i_l-N+1)} \dots A_{l(i_l-1)}\}$ and $\boldsymbol{B}_l = \{B_{l(i_l)} B_{l(i_l+1)} \dots B_{l(i_l+N-1)}\}$.

If $\boldsymbol{A}_l$ and $\boldsymbol{B}_l$ occur simultaneously with unit probability each, then jitter at i_l will never occur. However, some other events may be found, which do not obligatorily lead to jitter. We ignore such events and define approximately the probability $P(\boldsymbol{A}_l, \boldsymbol{B}_l)$ of the jitter-free breakpoint as

$$P(\boldsymbol{A}_l, \boldsymbol{B}_l) = P\left(\mathrm{A}_{i_l - N} \dots \mathrm{A}_{i_l - 1} \mathrm{B}_{i_l} \dots \mathrm{B}_{i_l + N - 1}\right). \tag{7}$$

The inverse event $\bar{P}(\boldsymbol{A}_l, \boldsymbol{B}_l) = 1 - P(\boldsymbol{A}_l, \boldsymbol{B}_l)$ can be called the *approximate jitter probability* [17].

2.3. Jitter distribution in the breakpoints

To determine the confidence limits for CNAs using high-resolution genomic arrays, jitter in the breakpoints must be specified statistically for the segmental Gaussian distribution. This can be done approximately if to employ either the discrete skew Laplace distribution or, more accurately, the modified Bessel function of the second kind and zeroth order.

2.3.1. Approximation with discrete skew Laplace distribution

Following the definition of the *jitter probability* given in Section 2.1 and taking into consideration that all the events are independent in white Gaussian noise, Eq. (7) can be rewritten as: $P(\boldsymbol{A}_l, \boldsymbol{B}_l) = P^N(A_l) P^N(B_l)$, where, following Eqs. (5) and (6), the probabilities $P(A_l)$ and $P(B_l)$ can be specified as, respectively,

$$P(A_l)=\begin{cases}1-\int_{\beta_l}^{\alpha_l} p_l(x)dx, & \sigma_l^2>\sigma_{l+1}^2,\\ \int_{\alpha_l}^{\infty} p_l(x)dx, & \sigma_l^2=\sigma_{l+1}^2,\\ \int_{\alpha_l}^{\beta_l} p_l(x)dx, & \sigma_l^2<\sigma_{l+1}^2,\end{cases} \tag{8}$$

$$P(B_l)=\begin{cases}\int_{\beta_l}^{\alpha_l} p_{l+1}(x)dx, & \sigma_l^2>\sigma_{l+1}^2,\\ \int_{-\infty}^{\alpha_l} p_{l+1}(x)dx, & \sigma_l^2=\sigma_{l+1}^2,\\ 1-\int_{\alpha_l}^{\beta_l} p_{l+1}(x)dx, & \sigma_l^2<\sigma_{l+1}^2,\end{cases} \tag{9}$$

where $p_l(x)=1/\sqrt{2\pi\sigma_l^2}e^{-((x-a_l)^2)/\sigma_l^2}$ is the Gaussian density.

Suppose that jitter occurs at some point $i_l \pm k,\ 0\le k\le N,$ as shown, for example, in **Figure 1**, and assign two additional blocks of events $\boldsymbol{A}_{lk}=\{A_{i_l-N}\cdots A_{i_l-1-k}\}$ and $\boldsymbol{B}_{lk}=\{B_{i_l+k}\cdots B_{i_l+N-1}\}$. The probability $P_k^-\triangleq P_k^-(\boldsymbol{A}_{lk}\bar{A}_{l(i_l-k)}\cdots\bar{A}_{i_l-1}\boldsymbol{B}_l)$ that jitter occurs at the kth point to the left from i_l (left jitter) and the probability $P_k^+\triangleq P_k^+(\boldsymbol{A}_l\bar{B}_{l(i_l+1)}\cdots\bar{B}_{l(i_l+k-1)}\boldsymbol{B}_k)$ that jitter occurs at the kth point to the right from i_l (right jitter) can thus be written as, respectively,

$$P_k^- = P^{N-k}(A_l)\left[1-P(A_l)\right]^k P^N(B_l), \tag{10}$$

$$P_k^+ = P^N(A_l)\left[1-P(B_l)\right]^k P^{N-k}(B_l), \tag{11}$$

By normalizing Eqs. (11) and (12) with Eq. (8), one can arrive at a function that turns out to be independent on N:

$$f_l(k)=\begin{cases}[P^{-1}(A_l)-1]^{|k|}, & k<0,\ (left)\\ 1 & k=0\\ [P^{-1}(B_l-1)]^k, & k<0.\ (right)\end{cases}. \tag{12}$$

Further normalization of $f_l(k)$ to have a unit area leads to the pdf $p_l(k)=\frac{1}{\varphi_l}f_l(k)$, where φ_l is the sum of $f_l(k)$ for all k,

$$\phi_l = 1 + \sum_{\infty}^{k=1} [\varphi_l^A(k) + \varphi_l^B(k)], \quad (13)$$

where $\varphi_l^A(k) = [P^{-1}(A_l) - 1]^k$ and $\varphi_l^B(k) = [P^{-1}(B_l) - 1]^k$.

It follows from the approximation admitted that $f_l(k)$ converges with k only if $0.5 < \tilde{P} = \{P(A),\ P(B)\} < 1$. Otherwise, if $\tilde{P} < 0.5$, the sum φ_l is infinite and $f_l(k)$ cannot be transformed to $p_l(k)$. It has been shown in [18] that such a situation is practically rare. It can be observed with extremely small and different segmental SNRs when the probabilities are comparable that the measurement point belongs to one of another segment.

Accepting $0.5 < \tilde{P} = \{P(A),\ P(B)\} < 1$, one concludes that $\tilde{P} < 0$, $\ln(1 - \tilde{P}) < 0$, and $\ln(1 - \tilde{P}) < \ln(\tilde{P})$. Next, using a standard relation $\sum_{k=1}^{\infty} x^k = 1 / (x^{-1} - 1)$, where $x < 1$, and after little transformations, Eq. (14) can be brought to

$$\phi_l = \frac{P(A_l) + P(B_l) - 1}{[1 - 2P(A_l)[1 - 2P(B_l)]}. \quad (14)$$

The *jitter pdf* $p_l(k)$ associated with the *lth* breakpoint can finally be found to be

$$p_l(k) = \frac{1}{\phi_l} \begin{cases} [P^{-1}(A_l) - 1]^{|k|}, & k < 0, \\ 1, & k = 0, \\ [P^{-1}(B_l) - 1]^k & k > 0, \end{cases} \quad (15)$$

where ϕ_l is specified by Eq. (15) and $0.5 < P(A_l),\ P(B_l) < 1$.

If now to substitute $q_l = P^{-1}(A_l) - 1$ and $d_l = P^{-1}(B_l) - 1$, find $P(A_l) = 1 / (1 + q_l)$ and $P(B_l) = 1 / (1 + d_l)$, and provide the transformations, then one may arrive at a conclusion that Eq. (16) is the discrete skew Laplace pdf [19].

$$p(k|d_l q_l) = \frac{(1 - d_l)(1 - q_l)}{1 - d_l q_l} \begin{cases} p_l^k, & k \geq 0 \\ q_l^{|k|} & k \leq 0 \end{cases}, \quad (16)$$

where $d_l = e^{-(\kappa_l / \nu_l)} \in [0,\ 1]$ and $q_l = e^{-(1/\kappa_l \nu_l)} \in [0,\ 1]$ and in which κ_l and $\nu_l > 0$ still need to be connected to Eq. (16). With this aim, consider Eqs. (16) and (17) at $k = -1$, $k = 0$, and $k = 1$. By equating Eqs. (16) and (17), first obtain $((1 - d_l)(1 - q_l) d_l) / (1 - d_l q_l) = 1 / \phi_l (1 - P(B_l)) / P(B_l)$ for $k = 1$ and $((1 - d_l)(1 - q_l) q_l) / (1 - d_l q_l) = 1 / \phi_l (1 - P(A_l)) / P(A_l)$ for $k = -1$ that yields

$$\nu_l = \frac{1-\kappa_l^2}{\kappa_l \ln \mu_l}, \tag{17}$$

where $\mu_l = (P(A_l)[1-P(B_l)])/(P(B_l)[1-P(A_l)])$. For $k=0$ we have $((1-d_l)(1-q_l))/(1-d_l q_l) = 1/\phi_l$ and transform it to the equation $x_l^2 - (\phi_l(1+\mu_l))/(1+\phi_l)x - (1-\phi_l)/(1+\phi_l)\mu_l = 0$, where a proper solution is

$$x = \frac{\phi_l(1+\mu_l)}{2(1+\phi_l)}\left(1-\sqrt{1-\frac{4\mu_l(1-\phi_l^2)}{\phi_l^2(1+\mu_l)^2}}\right) \tag{18}$$

and which $x_l = \mu_l^{-(\kappa_l^2)/(1-\kappa_l^2)}$ gives us

$$\kappa_l = \sqrt{\ln\frac{(x_l)}{\ln(x_l/\mu_l)}}. \tag{19}$$

By combining Eq. (18) with Eq. (20), one may also get a simpler form for ν_l, namely $\nu_l = -\kappa_l/\ln x_l$.

Now, introduce the segmental signal-to-noise ratios (SNRs): $\gamma_l^- = \frac{\Delta_l^2}{\sigma_l^2}$, and $\gamma_l^+ = \frac{\Delta_l^2}{\sigma_{l+1}^2}$, where $\Delta_l = a_{l+1} - a_l$, substitute the Gaussian pdf to Eqs. (9) and (10), provide the transformations, and rewrite Eqs. (9) and (10) as

$$P(A_l) = \begin{cases} 1+\frac{1}{2}\left[\mathrm{erf}\left(g_l^{\beta}\right)-\mathrm{erf}\left(g_l^{\alpha}\right)\right], & \gamma_l^- < \gamma_l^+ \\ \frac{1}{2}\mathrm{erfc}\left(g_l^{\alpha}\right), & \gamma_l^- = \gamma_l^+ \\ \frac{1}{2}\left[\mathrm{erf}\left(g_l^{\beta}\right)-\mathrm{erf}\left(g_l^{\alpha}\right)\right], & \gamma_l^- > \gamma_l^+ \end{cases}, \tag{20}$$

$$P(B_l) = \begin{cases} \frac{1}{2}\left[\mathrm{erf}\left(h_l^{\alpha}\right)-\mathrm{erf}\left(h_l^{\beta}\right)\right], & \gamma_l^- < \gamma_l^+ \\ 1-\frac{1}{2}\mathrm{erfc}\left(h_l^{\alpha}\right), & \gamma_l^- = \gamma_l^+ \\ 1+\frac{1}{2}\left[\mathrm{erf}\left(h_l^{\alpha}\right)-\mathrm{erf}\left(h_l^{\beta}\right)\right], & \gamma_l^- > \gamma_l^+ \end{cases}, \tag{21}$$

http://dx.doi.org/10.5772/63913

where $g_l^{\beta} = (\beta_l - \Delta_l) / |\Delta_l| \sqrt{\gamma_l^- / 2}$, $g_l^{\alpha} = (\alpha_l - \Delta_l) / |\Delta_l| \sqrt{\gamma_l^- / 2}$, $h_l^{\beta} = \beta_l / |\Delta_l| \sqrt{\gamma_l^+ / 2}$, $h_l^{\alpha} = \alpha_l / |\Delta_l| \sqrt{\gamma_l^+ / 2}$, $\mathrm{erf}(x)$ is the error function, $\mathrm{erfc}(x)$ is the complementary error function, and

$$\alpha_l, \beta_l = \frac{a_l \gamma_l^- - a_l \gamma_l^+}{\gamma_l^- - \gamma_l^+} \mp \frac{1}{\gamma_l^- - \gamma_l^+} \times \sqrt{(a_l - a_{l+1}) \gamma_l^- \gamma_l^+ + 2\Delta_l^2 (\gamma_l^- - \gamma_l^+) \ln \sqrt{\frac{\gamma_l^-}{\gamma_l^+}}} \ . \tag{22}$$

2.3.2. *Approximation of jitter distribution using the modified Bessel functions*

An analysis shows that the discrete skew Laplace pdf (17) gives good results only if SNR is >1. Otherwise, real measurements do not fit well, and a more accurate function is required. Below, we show that better approach to real jitter distribution can be provided using the modified Bessel functions.

2.3.2.1. *Modified Bessel function*

Figure 2 demonstrates the jitter pdf measured experimentally (dotted) for different SNRs. The breakpoint corresponds here to the peak density and the probability of the breakpoint location diminishes to the left and to the right of this point. Note that the discrete skew Laplace pdf (17) behaves linearly in such scales. Therefore, Eq. (17) cannot be applied when SNR is <1 and a more accurate function is required.

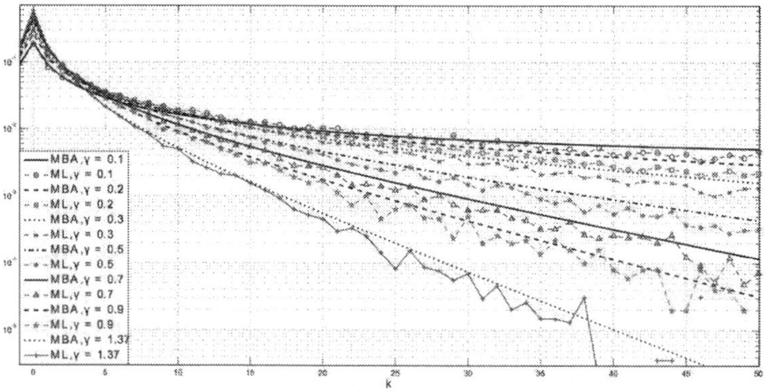

Figure 2. Experimentally defined one-sided jitter probability densities (dotted) of the breakpoint location for equal segmental SNR γ in the range of $M = 400$ points with a true breakpoint at $n = 200$. The experimental density functions were found using the *Maximum Likelihood* (ML) estimator. The histogram was plotted over 50×10^3 runs repeated nine times and average. Approximations (continuous) are provided using the proposed Bessel-based approximation depicted as MBA.

Among available functions demonstrating the pdf properties, the modified Bessel function of the second kind $K_0(x)$ and zeroth order is a most good candidate to fit the experimentally measured densities (**Figure 2**). The following form of $K_0(x)$ can be used:

$$
\begin{aligned}
K_0\left[x(k)\right] &= \int_0^{\infty} \cos\left[x(k)\sinh t\right] dt \\
&= \int_0^{\infty} \frac{\cos\left[x(k)t\right]}{\sqrt{t^2+1}} dt > 0, x(k) > 0 \,,
\end{aligned}
\tag{23}
$$

in which a variable $x(k)$ depends on index k, which represents a discrete departure from the assumed breakpoint location. Because $K_0[x(k)]$ is a positive-valued for $x(k)>0$ smooth function decreasing with x to zero, it can be used to approximate the probability density.

2.3.2.2. Approximation

In order to use Eq. (24) as an approximating function

$$
\boldsymbol{B}(k|\gamma) = K_0\left[x(k)\right] \tag{24}
$$

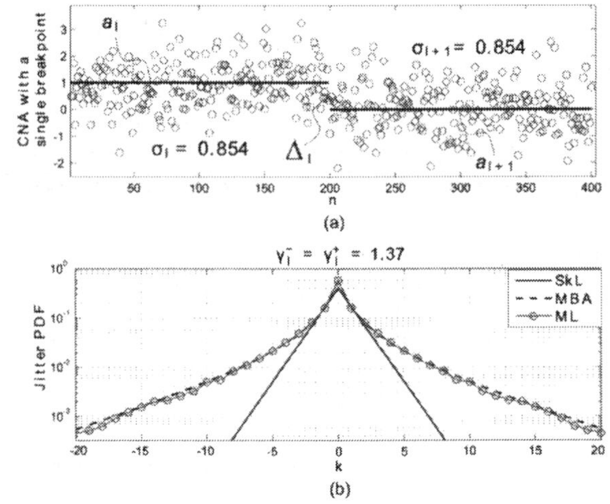

Figure 3. Simulated CNA with a single breakpoint at n = 200 and segmental standard deviations σ_l and σ_{l+1} corresponding to SNRs $\gamma_l^- = \gamma_l^+ = 1.37$: (a) measurement and (b) jitter distribution. Here, ML (circled) is the jitter pdf obtained experimentally using an ML estimator through a histogram over 50×10^3 runs, SkL (solid) is the Laplace distribution, and MBA (dashed) is the Bessel-based approximation.

conditioned on γ for the one-sided jitter probability densities shown in **Figure 2**, we represent a variable x via k as $x(k, \gamma)=\ln[\Phi(k, \gamma)]$ in a way such that small $k \geq 0$ correspond to large values x of and vice versa. Among several candidates, it has been found empirically that the following function $\Phi(k, \gamma)$ fits the histograms with highest accuracy:

$$\Phi(k,\gamma) = \left(|k|+1\right)^{\rho+\tau|k|}\left[\frac{1+\sqrt{\gamma}}{\gamma} - \epsilon\right], \tag{25}$$

if to set $\gamma = \gamma_l^-$ for $k<0$, $\gamma = \frac{\gamma_l^- + \gamma_l^+}{2}$ for $k=0$, and $\gamma = \gamma_l^+$ for $k>0$, and represent the coefficients and as $\tau(\gamma)$, $\rho(\gamma)$, and (γ) as

$$\tau(\gamma) = a_0\gamma + a_1 \tag{26}$$

$$\rho(\gamma) = \gamma\left(b_0\gamma^{b_1} + a_0\right) + b_2 \tag{27}$$

$$\epsilon(\gamma) = c_0\gamma^{c_1} + c_2 \tag{28}$$

where $a_0=0.02737$, $a_1=-4.5\times10^{-3}$, $b_0=0.3674$, $b_1=-0.3137$, $b_2=0.8066$, $c0=0.8865$, $c_1=-1.033$, and $c_2=-1.233$ were found in the mean square error (MSE) sense. These values were found in several iterations until the MSE reached a minimum.

In summary, **Figure 3** gives a typical example of a simulated CNA, where the modified Bessel function-based approximation (depicted as MBA) demonstrates better accuracy than the approximation obtained using the skew Laplace distribution (depicted as SkL).

2.4. Probabilistic masks

It follows from **Figure 3** that, in view of large noise, estimates of the CNAs may have low confidence, especially with small SNR $\gamma \leq 1$. Thus, each estimate requires confidence boundaries within which it may exist with a given probability [20, 21].

Given an estimate $\hat{a}_l$ of the lth segmental level in white Gaussian noise, the probabilistic upper boundary (UB) and lower boundary (LB) can be specified for the given confidence probability $P(\vartheta)$ in the ϑ-sigma sense as [20]

$$\hat{a}_l^{UB} \cong \hat{a}_l + \varepsilon = \hat{a}_l + \vartheta\sqrt{\frac{\sigma_j^2}{N_l}} = \hat{a}_l + \vartheta\hat{\sigma}_l \tag{29}$$

$$\hat{a}_l^{LB} \cong \hat{a}_l - \varepsilon = \hat{a}_l + \vartheta\sqrt{\frac{\sigma_j^2}{N_l}} = \hat{a}_l + \vartheta\hat{\sigma}_l \tag{30}$$

where ϑ indicates the boundary wideness in terms of the segmental noise variance $\hat{\sigma}_l$ on an interval N_l points, from $\hat{n}_{l-1}$ to $\hat{n}_l - 1$.

Likewise, detected the lth breakpoint location $\hat{n}_l$, the jitter probabilistic left boundary J_l^L and right boundary J_l^R can be defined, following [20], as

$$J_l^L \cong \hat{n}_l - k_l^R, \tag{31}$$

$$J_l^R \cong \hat{n}_l + k_l^L, \tag{32}$$

where $k_l^R(\vartheta)$ and $k_l^L(\vartheta)$ are specified by the jitter distribution in the ϑ-sigma sense.

By combining Eqs. (30) and (31) with Eqs. (32) and (33), the probabilistic masks can be formed as shown in [20] to bound the CNA estimates in the ϑ-sigma sense for the given confidence probability $P(\vartheta)$. An important property of these masks is that they can be used not only to bound the estimates and show their possible locations on a probabilistic field [20, 21] but also to remove supposedly wrong breakpoints. Such situations occur each time when the masks reveal double UB and LB uniformities in a gap of three neighboring detected breakpoints. If so, then the unlikely existing intermediate breakpoint ought to be removed.

Noticing that the segmental boundaries (30) and (31) remain the same irrespective of the jitter in the breakpoints, below we specify the masks for the jitter represented with the Laplace distribution (17) and Bessel-based approximation (25).

2.4.1. *Masks for Laplace distribution*

For the Laplace distribution (17), the jitter left boundary J_l^L (32) and right boundary J_l^R(33) can be defined in the ϑ--sigma sense if to specify $k_l^R(\vartheta)$ and $k_l^L(\vartheta)$ as shown in [18],

$$k_l^R = \left\lceil \frac{\nu}{\kappa} \ln \frac{(1-d_l)(1-q_l)}{\xi(1-d_l q_l)} \right\rceil, \tag{33}$$

$$k_l^L = \left\lceil \nu\kappa \ln \frac{(1-d_l)(1-q_l)}{\xi(1-d_l q_l)} \right\rceil, \tag{34}$$

where $[x]$ means a maximum integer lower than or equal to x. Note that functions (34) and (35) were obtained in [18] by equating (17) to $\xi(N_l)=\text{erfc}(\vartheta/\sqrt{2})$ and solving for k_l.

The probabilistic UB mask $\mathcal{L}_l^{UB}$ and LB mask $\mathcal{L}_l^{LB}$ for the Laplace distribution were formed in [17,20,21] by the segmental upper boundary $\hat{a}_l^{\text{UB}}$ and lower boundary $\hat{a}_l^{\text{LB}}$ and by the jitter left boundary J_l^{L} and jitter right boundary J_l^{R}. The algorithm for computing $\mathcal{L}_l^{UB}$ and $\mathcal{L}_l^{LB}$ masks has been developed and applied to the CNA probes in [22].

2.4.2. Masks for Bessel-based approximation

The UB mask $\mathcal{B}_l^{UB}$ and LB mask $\mathcal{B}_l^{LB}$ for the Bessel-based approximation can be formed using the same equations as for the Laplace distribution. Suppose that the Laplace pdf (17) is equal to the approximating function $\mathbf{B}_l(k)$ at $k=0$,

$$p\left(k=0|d_l,q_l\right)=\mathbf{B}_l\left(k=0\right), \tag{35}$$

that yields $\mathbf{B}_l(k=0)=\frac{1}{\phi_l}$. Then, define the probabilities $P^{B}(A_l)$ at $k=-1$ and $P^{B}(B_l)$ at $k=1$ as

$$p^{\mathbf{B}}\left(A_l\right)=\frac{\boldsymbol{B}_l\left(k=0\right)}{\mathbf{B}_l\left(k=-1\right)+\mathbf{B}_l\left(k=0\right)}, \tag{36}$$

$$p^{\mathbf{B}}\left(B_l\right)=\frac{\boldsymbol{B}_l\left(k=0\right)}{\mathbf{B}_l\left(k=1\right)+\mathbf{B}_l\left(k=0\right)}. \tag{37}$$

Next, substitute Eqs. (37) and (38) into Eqs. (19) and (20) to obtain κ_l^{B} and ν_l^{B}. The right-hand jitter k_l^{BR} and left-hand jitter k_l^{BL} can now be specified by, respectively,

$$k_l^{BR}=\left[\frac{\nu_l^{B}}{\kappa_l^{B}}\ln\frac{1}{\xi\boldsymbol{B}_l(k=0)}\right], \tag{38}$$

$$k_l^{BL}=\left[\nu_l^{B}\kappa_l^{B}\ln\frac{1}{\xi\boldsymbol{B}_l(k=0)}\right]. \tag{39}$$

Finally, define the jitter left boundary J_l^{BL} and right boundary J_l^{BR} as, respectively,

$$J_l^{BL}\cong\hat{n}_l-k_l^{BR}, \tag{40}$$

$$J_l^{BR} \cong \hat{n}_l - k_l^{BL}, \tag{41}$$

and use the algorithm previously designed in the study of Munoz-Minjares and Shmaliy [22] for the confidence masks based on the Laplace distribution.

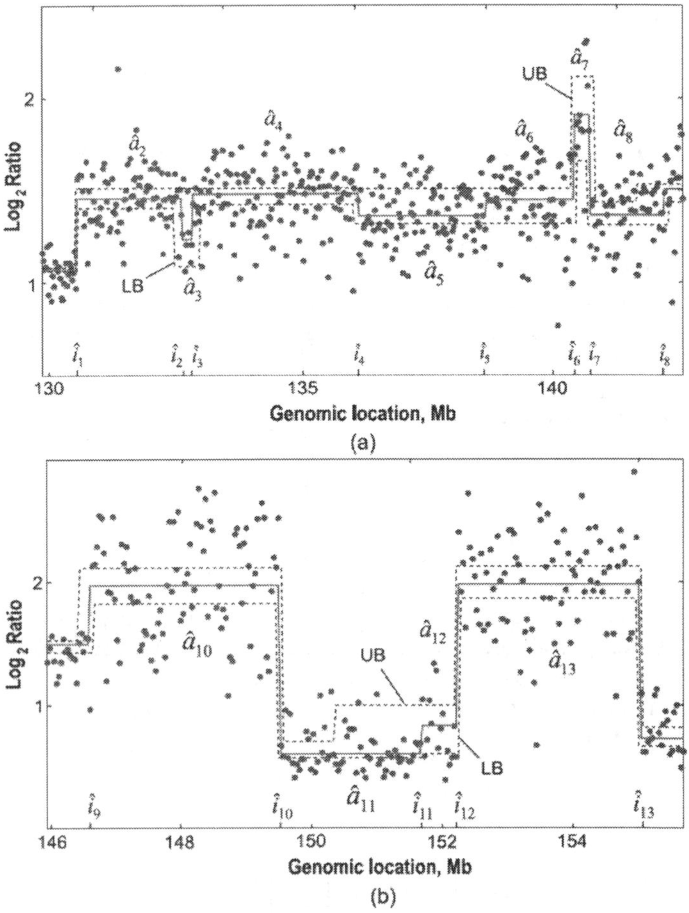

Figure 4. $\mathcal{L}_l^{UB}$ and $\mathcal{L}_l^{LB}$ masks for the seventh chromosome taken from "159A–vs–159D–cut" of ROMA: (a) genomic location from 130 to 146 Mb and (b) genomic location from 146 to 156 Mb. Breakpoints $\hat{i}_1$, $\hat{i}_6$, $\hat{i}_7$, $\hat{i}_9$, $\hat{i}_{10}$, $\hat{i}_{12}$, and $\hat{i}_{13}$ are well detectable because jitter is moderate. Owing to large jitter the breakpoints $\hat{i}_2$, $\hat{i}_3$, $\hat{i}_4$, $\hat{i}_5$, $\hat{i}_8$, $\hat{i}_9$, and $\hat{i}_{11}$ cannot be estimated correctly. There is a probability that the breakpoints $\hat{i}_2$, $\hat{i}_3$, $\hat{i}_4$, $\hat{i}_5$, and $\hat{i}_{11}$ do not exist. There is a high probability that breakpoint $\hat{i}_5$ does not exist.

3. Results

In this section, we test some CNA measurements and estimates by the algorithm developed in [22] based on the Laplace and Bessel approximations. In order to demonstrate the efficiency of the probabilistic masks and getting practically useful results, we exploit probes obtained by different technologies. First, we employ the results obtained with the HR-CGH profile and test them by the probabilistic masks using Laplace distribution. We next demonstrate the efficiency of the Bessel-based probabilistic masks versus the Laplace-based masks for the probes obtained with the SNP profile.

3.1. HR-CGH-based probing

The first test is conducted in the three-sigma sense suggesting that the CNAs exist between the UB and LB masks with high probability of $P = 99.73\%$. The tested HR-CGH array data are available from the representational oligonucleotide microarray analysis (ROMA) [23]. The breakpoint locations are also given in [23]. Voluntarily, we select data associated with potentially large jitter and large segmental errors. For clarity, we first compute some characteristics of the detected CNAs and notice that the segmental estimates found by averaging [18] are in a good correspondence with [23]. The database processed is a part of the seventh chromosome in archive "159A–vs–159D–cut" of ROMA a sample of B-cell chronic lymphocytic leukemia (B-CLL). It is shown to have 14 segments and 13 breakpoints (**Figure 4a** and **b**). Below, we shall show that, owing to large detection noise, there is a high probability that some breakpoints do not exist.

It follows from **Figure 4a** that the only breakpoint which location can be estimated with high accuracy is i_1. Jitter in i_6 and i_7 is moderate. All other breakpoints have large jitter. It is seen that the UB mask covering second to sixth segments is almost uniform. Thus, there is a probability that the second to fifth breakpoints do not exist. If to follow the LB mask, the locations of the second to fourth breakpoints can be predicted even with large errors. At least they can be supposed to exist. However, nothing definitive can be said about the fifth breakpoint and one may suppose that it does not exist. It is also hard to distinguish a true location of the eighth breakpoint. In **Figure 4b**, i_{10}, i_{12}, and i_{13} are well detectable owing to large segmental SNRs. The breakpoint i_9 has a moderate jitter. In turn, the location of i_{11} is unclear. Moreover, there is a probability that i_{11} does not exist.

3.2. SNP-based probing

Our purpose now is to apply the probabilistic mask with SNP profile that represents the CNA with low levels of SNR. Specifically, we employ the probes of the first chromosome available from "BLC_B1_T45.txt" a sample of primary breast carcinoma.

Inherently, the more accurate Bessel-based approximation extends the jitter probabilistic boundaries with respect to the Laplace-based ones, especially for low SNRs. We illustrate it in **Figure 5**, where the estimates of the first chromosome were tested by $\mathcal{B}_l^{UB}$, $\mathcal{B}_l^{LB}$, $\mathcal{L}_l^{UB}$, and $\mathcal{L}_l^{UB}$ for $\vartheta = 3$ (confidence probability $P = 99.73\%$).

In **Figure 6**, the masks $\mathcal{B}_l^{UB}$ and $\mathcal{B}_l^{LB}$ are placed in the vicinity of segment $\hat{a}_{18}$ for several confidence probabilities: $\vartheta = 0.6745 (P = 50\%)$, $\vartheta = 1 (P = 68.27\%)$, $\vartheta = 2 (P = 95.45\%)$, and $\vartheta = 3 (P = 99.73\%)$. What the masks suggest here is that the CNA evidently exists with high probability, but the segmental levels and the breakpoint locations cannot be estimated with high accuracy, owing to low SNRs.

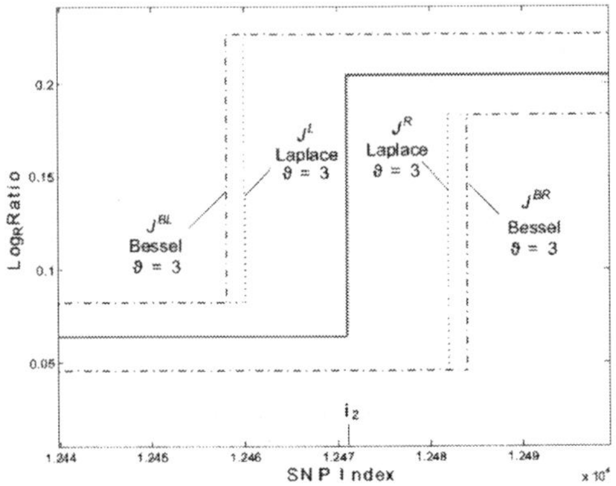

Figure 5. Jitter left boundaries $\mathcal{B}_l^{BL}$, J_l^{L} and right boundaries J_l^{BR}, J_l^{R} for the breakpoint $\hat{i}_2$ of first chromosome from sample BLC_B1_T45.txt (primary breast carcinoma). The probabilistic masks detect a breakpoint with a confidence probability $\vartheta = 3 (P = 99.73\%)$.

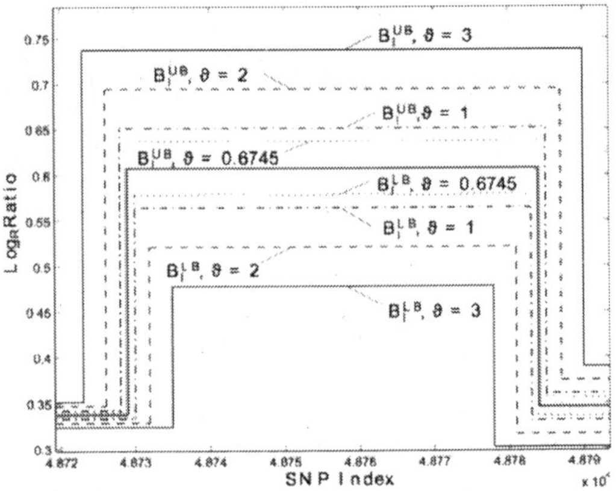

Figure 6. The $\mathcal{B}_l^{UB}$ and $\mathcal{B}_l^{LB}$ masks placed around the segmental level a_{18} for several confidence probabilities [20]. Here, the CNA exists with high probability, but the segmental levels and the breakpoint locations cannot estimate with high accuracy.

http://dx.doi.org/10.5772/63913

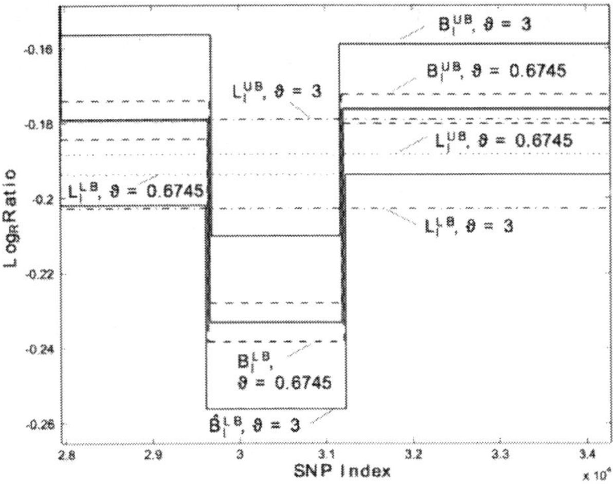

Figure 7. The confidence masks placed around a_{10} for $\vartheta=0.6745(P=50\%)$ and $\vartheta=3(P=99.73\%)$. Masks $\mathcal{L}_l^{UB}$ and $\mathcal{L}_l^{LB}$ do not confirm an existence of segmental changes while $\mathcal{B}_l^{UB}$ and $\mathcal{B}_l^{LB}$ indicate a small change.

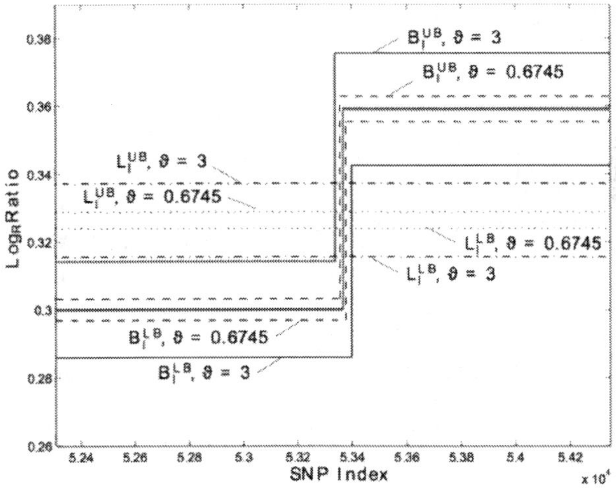

Figure 8. The confidence masks $\mathcal{L}_l^{UB}$, $\mathcal{L}_l^{LB}$, $\mathcal{B}_l^{UB}$ and $\mathcal{B}_l^{LB}$. placed around the breakpoint $\hat{i}_{20}$ for confidence probabilities $\vartheta=0.6745$ and $\vartheta=3$ of first chromosome from sample BLC_B1_T45.txt. The confidence masks based on Laplace distribution cannot detect the breakpoint $\hat{i}_{20}$.

A special case can also be noticed when the masks $\mathcal{L}_l^{UB}$ and $\mathcal{L}_l^{LB}$ are not able to confirm or deny an existence of segmental changes with high probability, owing to the inability of computing the Laplace-based masks for extremely low SNRs. **Figures 7** and **8** illustrate such situations. Just on the contrary, masks $\mathcal{B}_l^{UB}$ and $\mathcal{B}_l^{LB}$ can be computed for any reasonable SNR.

A conclusion that can be made based on the results illustrated in **Figures 5–8** is that the Bessel-based probabilistic masks can be used to improve estimates of the chromosomal changes for the required probability.

We finally notice that the computation time required by the masks to process the first chromosome from sample "BLC B1 T45.txt" with a length of n = 905215 was 2.634599 s using MATLAB software on a personal computer with a processor Intel Core i5, 2.5 GHz.

4. Discussion

We evaluate the breakpoints obtained by the projects representational oligonucleotide microarray analysis [23] and GAP [14] with the confidence masks. As has been shown before, not all of the detected chromosomal changes have the same confidence to mean that there is a probability that some breakpoints do not exist. In order to improve the CNA estimates for the required confidence, the following process can be used:

1. Obtain estimates of the CNA using the standard CBS algorithm [24, 25] or any other algorithm.

2. Compute masks $\mathcal{B}_l^{UB}$ and $\mathcal{B}_l^{LB}$ for the given confidence probability P, % and bound the estimates.

3. If the masks reveal double uniformities, in UB and LB, in a GAP of any three neighboring breakpoints, then remove the intermediate breakpoint and estimate the segmental level between the survived breakpoints by simple averaging. The CNAs estimated in such a way will be valid for the given confidence P, %.

Application of this methodology to the CNA structure detected in frames of the Project GAP is shown in **Figure 9**. Its special feature is a number of hardly recognized small chromosomal changes (**Figure 9a**). We test them by the proposed masks $\mathcal{B}_l^{\mathrm{UB}}$ and $\mathcal{B}_l^{\mathrm{LB}}$. To this end, we first start with equal confidence probabilities of P=50% for each estimate to exist or not exist and find out that three breakpoints demonstrate no detectability. We remove these breakpoints and depict their locations with "×". Reasoning similarly, we remove four breakpoints to retain only probable changes, by P=75%, nine breakpoints to show a picture combined with almost certain changes, by P=93%, and 10 breakpoints in the three-sigma sense, P=99.73%. Observing the results, we infer that the masks are able to correct only the estimates obtained under the low SNRs. The relevant chromosomal sections S1–S7 are circled in **Figure 9**. It is not surprising because changes existing with high SNRs are seen visually. An estimator thus can easily detect them with high confidence.

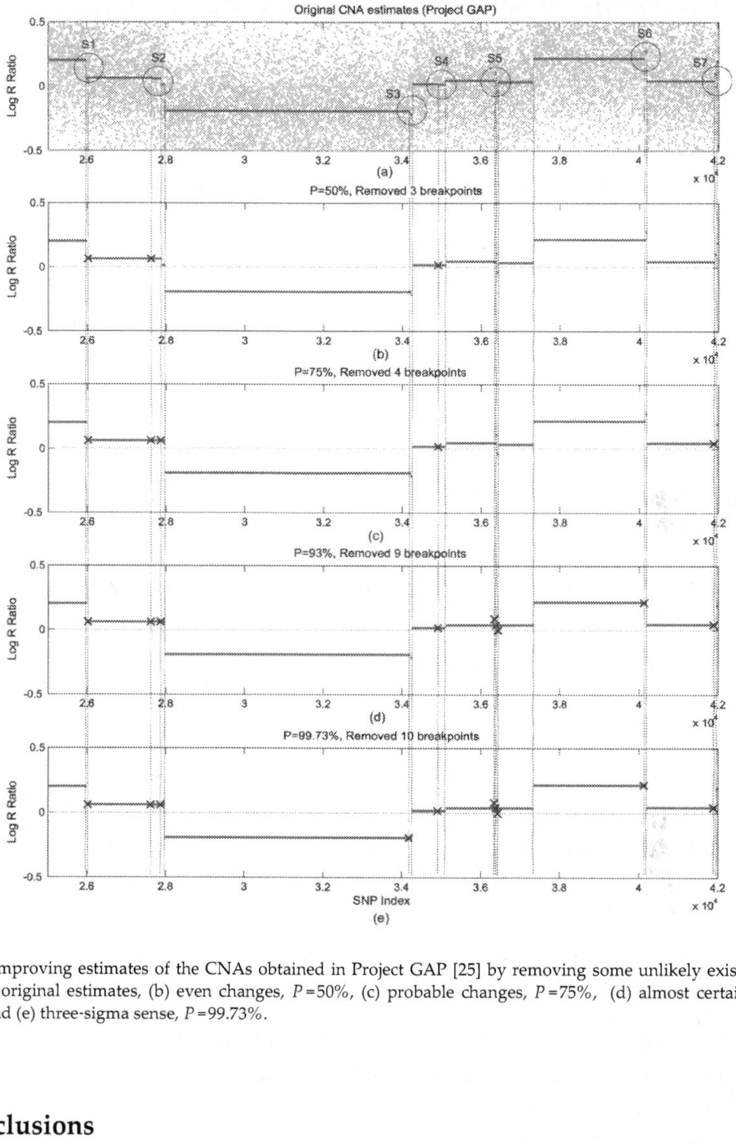

Figure 9. Improving estimates of the CNAs obtained in Project GAP [25] by removing some unlikely existing breakpoints: (a) original estimates, (b) even changes, $P=50\%$, (c) probable changes, $P=75\%$, (d) almost certain changes, $P=93\%$, and (e) three-sigma sense, $P=99.73\%$.

5. Conclusions

Modern technologies developed to produce the CNA profiles with high resolution still admit intensive white Gaussian noise. Accordingly, not one estimator even ideal is able to provide jitter-free estimation of segmental changes. Thus, in order to avoid wrong decisions, the estimates must be bounded for the confidence probability. Jitter exists in the CNA's breakpoints fundamentally. When SNR is >1, it can statistically be described using the discrete skew Laplace distribution. Otherwise, if SNR is <1, the Bessel-based approximation produces more accuracy. By the jitter distribution, it is easy to find a region within which the breakpoint exists for the required probability. Of practical importance are the confidence UB and LB masks, which can be created based on the segmental and jitter distributions for the given confidence probability. The masks can serve as an auxiliary tool for medical experts to make decisions about the CNA structures. Applications to probes obtained using the HR-CGH and SNP technologies confirm efficiency of the confidence masks.

Author details

Jorge Muñoz-Minjares*, Yuriy Shmaliy and Oscar Ibarra-Manzano

*Address all correspondence to: ju.munozminjares@ugto.mx

Department of Electronics Engineering, University of Guanajuato, Salamanca, Mexico.

References

[1] Reymond, A., Henrichsen, C. N., Harewood, L., and Merla, G. Side effects of genome structural changes. Curr Opin Genet Dev. 2007;17(5):381–386.

[2] Alkan, C., Coe, B. P., and Eichler, E. E. Genome structural variation discovery and genotyping. Nat Rev Genet. 2011;12(5):363–376.

[3] Feuk, L., Carson, A. R., and Scherer, S. W. Structural variation in the human genome. Nat Rev Genet. 2006;7(2):85–97.

[4] Redon, R., Ishikawa, S., Fitch, K. R., Feuk, L., *et al*. Global variation in copy number in the human genome. Nature. 2006;444(7118):444–454.

[5] Hastings, P. J., Lupski, J. R., Rosenberg, S. M. and Ira, G. Mechanisms of change in gene copy number. Nat Rev Genet. 2009; 10 (8): 551–564.

[6] Conrad, D. F., Pinto, D., Redon, R., Feuk, L., Gokcumen, O. *et al*. Origins and functional impact of copy number variation in the human genome. Nature. 2010;464(7289): 704–712.

[7] International Human Genome Sequencing Consortium. Finishing the euchromatic sequence of the human genome. Nature. 2004;431(7011): 931–945.

[8] Forozan, F., Karhu, R., Kononen, J., Kallioniemi, A., and Kallioniemi, O. P. Genome screening by comparative genomic hybridization. Trends in Genetics. 1997; 13 (10): 405– 409.

[9] Speicher, M. R. and Carter, N. P. The new cytogenetics: blurring the boundaries with molecular biology. Nat Rev Genet.2015;6(10): 782–792.

[10] Ng, P. C. and Kirkness, E. F. Whole genome sequencing. Methods MolBiol. 2010;628: 215–226.

[11] Wang, D. G., Fan, J. B., Siao, C. J., Bermo, A., Young, P., *et al.* Large-scale identification, mapping, and genotyping of single-nucleotide polymorphisms in the human genome. Science. 1998;280 (5366): 1077–1082.

[12] Yang, Y. H., Dudoit, S., Luu, P., Lin, D. M., et al. Normalization for cDNA microarray data: a robust composite method addressing single and multiple slide systematic variation. Nucleic Acids Res. 2002;30(4): 1–10.

[13] Urban, A. E., Korbel, J. O., Selzer, R., Richmond, T., Hacker, A., Popescu, G. V., Cubells, J. F., Green, R., Emanuel, B. S., Gerstein, M.B., Weissman, S. M., Snyder, M. High-resolution mapping of DNA copy alterations inhuman chromosome 22 using high-density tiling oligonucleotide arrays. Proc Natl Acad Sci U S A. 2006;103(12): 4534–4539.

[14] Popova, T., Boeva, V., Manie, E., Rozenholc, Y., Barillot, E., and Stern, M.H. Analysis of Somatic Alterations in Cancer Genome: From SNP Arrays to Next Generation Sequencing. Genomics I Humans, Animals and Plants, 2013. <hal-01108425> Observation to publisher: Reference from https://hal.archives-ouvertes.fr/hal-01108425

[15] Pique-Regi, R., Ortega, A., Tewfik, A., and Asgharzadeh, S. Detection changes in the DNA copy number. IEEE Signal Process Mag. 2012;29(1): 98–107.

[16] Munoz-Minjares, J., Cabal-Aragón, J., and Shmaliy, Y.S. Jitter probability in the breakpoints of discrete sparse piecewise-constant signals. Proc. 21st Eur. Signal Process. Conf. (EUSIPCO-2013), Marrakech, Marocco. 2013; pp. 1–5.

[17] Munoz-Minjares, J., Shmaliy, Y. S., and Cabal, A. J. Noise studies in measurements and estimates of stepwise changes in genome DNA chromosomal structures. Adv. App Pure Math. 2014.

[18] Munoz-Minjares, J., Cabal-Aragón, J., and Shmaliy, Y. S. Probabilistic bounds for estimates of genome DNA copy number variations using HR-CGH microarray. Proc. 21st European Signal Process. Conf. (EUSIPCO-2013). 2013; pp. 1–5.

[19] Kozubowski, T. J. and Inusah, S. A skew Laplace distribution on integers. Ann Inst Stat Math 2006;58(3):555–571.

[20] Munoz-Minjares, J., Cabal-Aragon, J., and Shmaliy Y. S. Confidence masks for genome DNA copy number variations in applications to HR–CGH array measurements. Biomed Signal Process Contr. 2014;13:337–344.

[21] Munoz-Minjares, J., Shmaliy, Y. S., and Cabal-Aragon, J. Confidence limits for genome DNA copy number variations in HR–CGH array measurements. Biomed Signal Process Contr. 2014;10:166–173.

[22] Munoz-Minjares, J. and Shmaliy, Y. S.: Bounding error in estimates of genome copy number variations using SNP array. International Journal of Biomedical Engineering. 2015;9:127–132.

[23] Lucito, R., Healy, J., Alexander, J., Reiner, A., Esposito, D., Chi, M., Rodgers, L., Brady, A., Sebat, J., Troge, J., West, J. A., Rostan, S., Nguyen, K. C. Q., Powers, S., Ye, K. Q., Olshen, A., Venkatraman, E., Norton, L., Wigler, M. Representational oligonucleotide microarray analysis: a high-resolution method to detect genome copy number variation. Genome Res. 2003;13(10): 2291–2305.

[24] Olshen, A. B., Venkatraman, E. S., Lucito, R., and Wigler, M. Circular binary segmentation for the analysis of arraybased DNA copy number data. Biostatistics. 2004;5(4): 557–572.

[25] Venkatraman E. S. and Olshen, A. B. A faster circular binary segmentation algorithm for the analysis of array CGH data. Bioinformatics. 2007;23(6):657–663.